Design at Work:

Cooperative Design of Computer Systems

Design at Work:

Cooperative Design of Computer Systems

Edited by

JOAN GREENBAUM, City University of New York
MORTEN KYNG, Aarhus University, Denmark

CRC Press
Taylor & Francis Group
Boca Raton London New York

CRC Press is an imprint of the
Taylor & Francis Group, an **informa** business

Graphic Design:
Morten Kyng, Karen Kjær Møller, and Joan Greenbaum

Illustrations:
Gail Mardfin Strakey and Steve Wallgren (p. 66f)

First Published by
Lawrence Erlbaum Associates, Inc., Publishers
365 Broadway
Hillsdale, New Jersey 07642

6000 Broken Sound Parkway, NW Suite 300, Boca Raton, FL 33487
270 Madison Ave, New York, NY 10016
2 Park Square, Milton Park, Abingdon, Oxon OX14 4RN, UK

Library of Congress Cataloging-in-Publication Data

Design at work : cooperative design of computer systems / edited by
Joan Greenbaum, Morten Kyng.
        p.   cm.
    Includes bibliographical references and index.
        ISBN 0-8058-0611-3 — ISBN 0-8058-0612-1 (pbk.)
        1. System design. I. Greenbaum, Joan M., 1942- . II. Kyng,
    Morten, 1950.
    QA76.9.S88D473 1991
    004.2'1--dc20   90-3942
                                                 CIP

Publisher's Note
The publisher has gone to great lengths to ensure the quality of this reprint
but points out that some imperfections in the original may be apparent.

CONTENTS

Preface: Memories of the Past

Joan Greenbaum and Morten Kyng

In the Autumn of 1984, my colleague Pelle Ehn and I went on a trip to the United States to discuss our experiences from Denmark and Sweden with involving workers in the design of computer systems. By then I had been working together with "end users" and trade unions for a decade, but I was still considered an outsider at my university: Working with end users was not "academic," and doing design instead of controlled experiments was considered "unscientific." The people we visited in the U.S. were not mainstream computer scientists, either. But the interest we encountered, as well as the discussions we had, convinced us that designing cooperatively with end users was worth pursuing, not only in a Scandinavian setting. It had something to offer system designers who wanted computers to improve people's work, and it had a lot to offer end users, who generally don't get the chance to be actively involved in the design process.

Morten Kyng

In the Fall of 1986, I was invited to Aarhus University, Denmark to teach a course on system development. Packed deep in my baggage, along with the chaos of traveling with my children and my dog, was the fear of not finding a textbook that could begin to explain the excitement of creating computer systems. This wasn't a new fear. Like all teachers of the art and would-be science of system design know, this problem unwraps itself every semester. And having worked for years as a computer system consultant, I knew that the endless range of "how-to" books could never fill the void between the rich experiences of practitioners and the inquisitive world of students eager to begin building their own experiences.

After unpacking my suitcases, sending my children off to school equipped with a hastily learned word or two of Danish, and arranging and re-arranging my office, my colleague Morten Kyng and I began a series of discussions about our past experiences and about what we hoped to capture in a new course. The environment was ripe. Aarhus was one of those points in time and space where a critical mass of computer and social scientists were asking each other questions.

Joan Greenbaum

We wish we could say, as many authors do, that the idea of this book took shape at that time. However, there were several years where we were to muddle through the intricacies of teaching, experimenting, discussing, and expanding our circle of colleagues, before we could catch our collective breath and reflect on the process we had begun. Identifying the problems with our past design experiences was comparatively easy. Giving shape to our visions of the future was so exciting it barely passed for work. But as everyone who has ever grappled with the design process knows, addressing that perennial question "How do we make it happen?" is excruciatingly difficult.

From our perspective it is now clear that system development is difficult, not because of the complexity of technical problems, but because of the social interaction when users and system developers learn to create, develop and express their ideas and visions. Traditional methods and practices are firmly rooted in the engineering/natural science origins of computer science. They are designed for implementing clear-cut specifications (as if such things exist), but they are far removed from actual practice. In fact, traditional methods, with their emphasis on step-by-step procedures, effectively prevent creative and cooperative sparks between system designers and users.

This book then is a cooperative attempt to ignite the human sparks of imagination and creativity, so that they can burn themselves into useful computer systems. After all, a computer system is not merely an assembly of silicon chips ordered to solve a particular problem. As all who have used computers in the workplace or at home know, their usefulness depends on the fragile relationship of the person, the working environment, and the computer technology itself. We call our approach cooperative design because it emphasizes designers and users working actively together, and because it supports the collaboration of people from different disciplines looking at workplace activities from multiple perspectives.

The experiences of many of the authors in this book come from a tradition that is sometimes referred to as the Scandinavian approach to system design. This approach, begun in the 1970s with Norwegian trade union-supported projects, has as its focal idea the involvement of workers, as users of technology, in the design of the tools that they are using in their workplace. Today, the idea of user-participation is no longer new, but the experiences gained from the early Scandinavian projects, as well as their overall philosophy that emphasizes empowering workers, distinguish the Scandinavian approach from the traditional approaches to user involvement. A driving force behind this book is our interest in highlighting key ideas in both Scandinavian and American participatory design philosophies, and building a bridge that supports and promotes users interests. For we see cooperative system design as more than props or background to create "user friendly" systems. Rather, we see the need for users to become full partners in a cooperative system design process where the pursuit of users' interest is a legitimate element.

This book brings people from the humanities, social sciences, and computer science together at a time when the borders between the disciplines have, thankfully, we believe, been collapsing. No longer are we viewed as heretics for stepping outside of our respective professions and trying to borrow ideas and methods from our colleagues in other areas. The authors selected have been included because they have written or worked on ideas that have challenged the boundaries of system design and stirred the imagination of students who have learned about them.

We've undertaken this writing project as we would a system development project—as a cooperative endeavor. Attempting to apply the ideas that we write about has not been without its conflicts. As the chapter references suggest, what is left out of the book is a huge and significant world of activities that we unfortunately couldn't begin to tackle in this project. We would like you to think about our efforts as a collage of ideas to reflect on and experiment with. Perhaps a video of our collective experiences could more accurately capture the spirit of what we are trying to do. But the video will have to wait. Bring your imagination, instead, into these chapters, and join us.

Acknowledgments

Over a period of two years the authors of this book held several meetings where we read and reread each others' contributions. We particularly want to thank each of the contributors for their ideas and efforts in making it all come together. We want to thank Julia Hough whose encouragement got us started and our editor, Hollis Heimbouch, for her continued support and encouragement. We are especially grateful to Karen Kjær Møller whose truly unending work has coordinated all of our efforts, to Gail Mardfin Starkey whose imagination brought forth the drawings for this book, and to Jonathan Grudin whose insight and careful attention to detail helped us pull this book together. We want to thank the Computer Science Department at Aarhus University for its support as "host" for the production of the manuscript, and we are grateful to our children Bart and Jesse Greenbaum and Kasper and Rasmus Kyng for being patient with us while we were distracted in the world of "Design at Work."

Joan Greenbaum   *Morten Kyng*
Montclair, New Jersey   *Aarhus, Denmark*

1

Introduction: Situated Design

Joan Greenbaum and Morten Kyng

Well, I don't really know how to tell you what I don't like about the system. I guess one of the things is that it makes me think and work differently, like for example, when I want to make separate columns, I need to type it and then rearrange it. That's not the way I see it in my mind.

Word processing user

If a lion could speak would we understand her?

Paraphrase of Wittgenstein, 1953, p. 223

As the dust of the 1980s settles, we who design computer systems for the workplace know a little more about working with the people who will use them. But just a little. Our intention in writing this book is to document some of the things we know, and to point to a growing body of projects where the people who use computers and the people who design them have learned more about each other's work and about what happens in their workplaces.

In 1985 in Aarhus, Denmark, several of the contributors to this book organized a conference on Computers and Democracy (Bjerknes, Ehn, & Kyng, 1987). That conference, a follow-up to over a decade of involvement in Scandinavian user centered design projects, brought out a kaleidoscope of visions evolving from the original idea of workers-as-user participant. At the risk of confining the wide range of visions, the authors in this book share, as general background, a set of design ideals which have guided our work. These are:

- Computer systems that are created for the workplace need to be designed with *full participation* from the users. Full participation, of course, requires training and active cooperation, not just token representation in meetings or on committees.
- When computer systems are brought into a workplace, they should *enhance* workplace skills rather than degrade or rationalize

them. Enhancing skills means paying attention to things that are often left out of the formal specifications, like respect for tacit knowledge, building on shared knowledge and, most importantly, communication. Computer systems are a lot more than the simple flow of information represented in the flowcharts that systems analysts present to their clients.

- Computers systems are *tools*, and need to be designed to be under the control of the people using them. They should support work activities, not make them more rigid or rationalized.

- Although computer systems are generally acquired to increase productivity, they also need to be looked at as a means to increase the *quality* of the results. More output, such as the reams of printed pages emerging from many management information systems, doesn't mean better output. The double emphasis on productivity and quality raises new questions in the design process (see Daressa, 1986).

- The design process is a political one and includes *conflicts* at almost every step of the way. Managers who order the system may be at odds with workers who are going to use it. Different groups of users will need different things from the system, and system designers often represent their own interests. Conflicts are inherent in the process. If they are pushed to the side or ignored in the rush to come up with an immediately workable solution, that system may be dramatically less useful and continue to create problems.

- And finally, the design process highlights the issue of how computers are used in the context of work organization. We see this question of focusing on how computers are used, which we call the *use situation,* as a fundamental starting point for the design process. In fact, it is here that we put our attention in the first part of the book by introducing ideas about how people work.

These ideas were a base for the authors as we began new projects in the late 1980s. Our experiences in these projects have shaped an emerging approach to cooperative or participatory design that focuses on workplace activities. Part I, Reflecting on Work Practice, sketches our understanding of how people work together and what we need to do when we look at the intricate fabric of workplace activity. Part II, Designing for Work Practice, tells a number of different stories from the on-going "dialogue" between a wide variety of our empirical design projects and our reflections on the process of design.

Users as Competent Practitioners

To system designers, the people who use computers are awkwardly called "users," a muddy term that unfortunately tends to focus on the people sitting in front of a screen rather than on the actual work people are doing. The word lumps all kinds of workplace activity together implicitly putting the computer in focus and treating people as a blurred background. Like Wittgenstein's riddle about the lion, these users are all too often understood by system developers in "system terms." Just as the human observer misleadingly assigns meaning to what lions are doing based on the human's own world view, system developers tend to make sense out of the work of the users by applying their own system development concepts, often missing the understanding of the users which stems from a knowledge of and experience with the work being done. Wittgenstein's point in the lion riddle is that understanding between humans and lions is not possible because they don't share a common practice. Fortunately, we believe our possibilities for mutual understanding with users are much better. However, it is not something which is there, a priori, but something that has to be carefully developed.

Although the identification of user issues now dominates the computer management and system development literature, the majority of books, articles, and seminars addresses the issue of how best to "integrate the user" into the system development process (see *Communications of the ACM*, Jan. 1990). We set ourselves apart from this, for the intent of this book is to show, through examples, how the diverse groups of people called users can actively learn, participate, and cooperate with system designers. Our interest is not in fitting users into an already existing system development process, but in creating new ways of working together. For us, user participation does not mean interviewing a sample of potential users or getting them to rubber stamp a set of system specifications. It is, rather, the active involvement of users in the creative process we call design.

In this book you will meet a wide range of users—people whose jobs include baggage handler, dental assistant, clerical worker, journalist, librarian, and typographer. They are some of the people with whom we, as authors, computer scientists, systems designers, linguists, and social scientists have worked. We see users not as one homogeneous group, but, rather, as diverse groups of people who have competence in their work practices. Our perspective focuses explicitly on all the different groups of people using computers in their work, and not on the managers. There has been a great deal written about management objectives in the system development process. We shift that focus and work directly with the people whom system developers typically call "end-users."

By viewing the people who use computers as competent in their field of work, we find that the workplace takes on the appearance of a rich tapestry, deeply woven with much intricacy and skill. The first part of the book, Reflecting on Work Practice, examines this tapestry and finds four underlying patterns. These are:

- the need for designers to *take work practice seriously* ;
- the fact that we are dealing with *human actors*, not cut-and-dried human factors;
- the idea that work tasks must be seen within their context, and are therefore *situated actions;* and perhaps the most important of all, as it links the rest together,
- that *work is fundamentally social*, involving extensive cooperation and communication.

This last pattern, the focus on the social nature of work practice, makes us acutely aware of the *cooperative nature* of workplace tasks. Our interest in cooperative activities does not mean that we glorify the way people work with each other, but rather it forces us to look at the ways people in an organization create, use, and change information, knowledge, and tasks. Few work tasks are done in isolation, and fewer still are easy to describe. While traditional system development methods treat specific work tasks as formalizable data processing done by individuals in isolation, communicating via data channels, the premise of our approach shifts to looking at groups interacting in multifarious ways within complex organizational contexts. And, by the same token, although traditional methods require describing these work tasks, our approach is based on the belief that the complex pattern of workplace life *is not easily describable*. Therefore, we need new tools and techniques to capture this complexity, and to develop a more detailed understanding of its depth.

The second part of the book, Designing for Work Practice, takes a look at the kinds of techniques designers and users can employ to bring this seemingly indescribable rich tapestry to bear on design. Here we begin with the idea that each of us understands the workplace from within our own experience. Thus the computer system designer, peering into a workplace from the outside, can't expect to capture the same meanings that someone involved in day to day activity could. And the designers cannot automatically expect the users to participate creatively in design activities, which may be completely new to them. Again bringing the lion analogy back into the picture, since neither designer nor user groups can *fully* understand each others' practices or meanings, we need to build a bridge that brings these experiences closer together. Part II addresses ways to bridge this gulf, presenting examples from case studies in order to provide an approach based on *cooperative action* rather than formal description. The underlying ideas in this section involve:

- *mutual learning* between users and designers about their respective fields (see also Bjerknes & Bratteteig, 1984);
- use of tools in the design process that are *familiar* to the users;
- *envisionment* of future work situations to allow the users to experience how emerging designs may affect the work practice rather than relying on the seemingly esoteric language of system developers; and
- the importance of starting the design process *in the practice of the users*.

The traditional scales of system development have placed far more weight on getting users to understand the language of system designers. The authors in this part present their experiences as a way of creating room for users to act, and thus we hope to balance the scales.

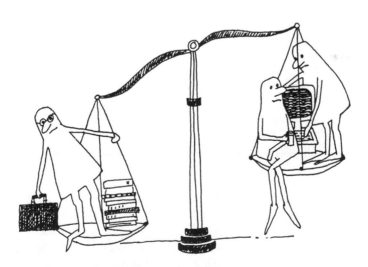

This book, then, is an attempt to look at the development of computer systems as a process, where people as living, acting beings, are put back into the center of the picture. A picture of the workplace, where the situations people find themselves in, with all of its conflict-laden social and political tensions, comes under close scrutiny. The authors have been invited to write chapters on the basis of their experiences in undertaking projects that attempt to do this. They are not in agreement on the question of how computer system designers *should* do things, but, as we mentioned in the Preface, this is not another how-to book. They are in agreement, however, on the fact that design needs to take place with users as full and active participants, and that the tools and techniques for doing this are dependent on the situations within the workplace.

Our focus on starting design with our eyes firmly fixed on the interactions taking place within the workplace was inspired, in part, by Lucy Suchman's book, *Plans and Situated Actions* (1987). Her work takes as its starting point the idea that human actions are not so much guided by concrete plans as based on situations. Thus, as the circumstances in which we find ourselves change, so do our actions. Most books on system design begin with the need for system designers to carefully define the "problem." This book doesn't. Instead we ask ourselves and our readers to consider design issues from the broader context of the *situated actions* of the people with whom they are working. This chapter, which serves as an introduction to the book, takes a look at the changing circumstances that led us to this perspective.

As the book title states, our approaches are based on cooperation between system developers and those people we call users. But it also implies that most work is cooperative and that the process of putting this book together, like any collaborative venture, involved a great deal of interaction among people of different disciplines. The theme of cooperation or respect for *mutual competencies*, whether they be between designers and users, or authors in this book, is a central one for us. Just as we see users as diverse groups of competent practitioners, we have had to look at ourselves, as authors, as a diverse assortment of academic practitioners who speak different professional languages and use different approaches. We are lucky to be writing this at a time when walls between academic fields are beginning to collapse. In fact, by virtue of having undertaken projects where we looked at workplaces from a variety of perspectives and designed systems with people who use them, we have contributed to collapsing these boundaries. We hope to do more.

At the end of this chapter we will return to a discussion of the background of the authors and their contributions to the book. It is time now to take a look at how we got here, by examining some of the history of things that have gone wrong in developing computer systems, and then, of course, by looking at some of the more optimistic trends that have led us to a better understanding of Design at Work. We had originally planned to call the book "Design by Doing," highlighting our focus on "doing" work cooperatively with users, and on using methods that emphasize action more than formal description. But in the process of finding out what we were doing individually and by sharing our experiences, we found that before we could actually do "design by doing," we had to develop a more in-depth picture of what takes place at work and what practices constitute cooperative design. Thus, *Design at Work* brings to the foreground our enthusiasm for the work of design and for the importance of designing to support what actually takes place at work.

What's Gone Wrong?

Writing in 1965, Robert Boguslaw attempted to challenge computer system developers, whom he called "the new utopians," by saying:

> And so it is that the new utopians retain their aloofness from human and social problems presented by the fact or threat of machined systems and automation. They are concerned with neither souls nor stomachs. People problems are left to the after-the-fact efforts of social scientists. (p. 3)

Twenty-five years after Boguslaw's warning, and the warnings of many others (see Ackoff, 1974; Hoos, 1961; Nygaard & Bergo, 1975; Weizenbaum, 1976), we are repeatedly "surprised" by computer systems that don't work as intended. In almost every workplace there are stories that pass from department to department about computer systems that simply do not work. Often the stories take the form of legend, as workers tell the tale of how they "got around some dumb" computer system. As system developers, we like to think that we have learned a lot from our overzealous 'mistakes' of the past. But from both computer system literature and workplace folklore, the fact remains that many computer systems don't fit the work activities of the people who use them.

Thinking of the past as some random pattern of mistakes will not help get us away from being stuck in some blind alleys. If we were to look, instead, at the history of western scientific thought, we would find that the objectives and methods of computer systems development clearly grow out of several centuries of beliefs and practices in the natural sciences. According to Terry Winograd and Fernando Flores (1986), these deeply rooted practices, referred to as the rationalistic tradition, do the following:

1. Characterize the situation in terms of identifiable objects with well-defined properties.
2. Find general rules that apply to situations in terms of those objects and properties.
3. Apply the rules logically to the situation of concern, drawing conclusions about what should be done. (p. 15)

These step-by-step, rather linear processes force our thinking into narrow pathways, where emphasis is placed on isolating single problems and searching for the one "right" solution. While these ideals may be suited to certain types of scientific problem-solving, we feel that they are ill-suited to the dynamic and generally chaotic conditions of developing computer systems for the workplace. As Boguslaw pointed out, they channel our thinking away from "souls and stomachs."

This rationalistic tradition has its roots in Cartesian philosophy, which tends to split the world into the inner world of our minds and the outer world of things, thus dividing mind and body and drawing a line between our emotional interior and the objective, thing-oriented environment. In *Work-Oriented Design of Computer Artifacts*, Pelle Ehn (1989) tells us how this applies to system development:

> The prototypical Cartesian scientist or system designer is an observer. He does not participate in the world he is studying, but goes home to find the truth about it by deduction from the objective facts that he has gathered. (p. 52)

Although some might think that Ehn's example seems like a caricature of the way we work, a look at system development literature will verify his point. The dominant threads in systems literature today, reflected in the well-read works of DeMarco (1978), Yourdon (1986), and Jackson (1983), illustrate the call to objectivity and problem isolation. Edward Yourdon, in one of his many books on the system approach, defines his design strategy as one "that breaks large complex problems into smaller less complex problems and then decomposes each of these smaller problems into even smaller problems, until the original problem has been expressed as some combination of many small solvable problems" (p. 61).

Of course day-to-day systems practice is different from the formal methodology that traditional systems people, like Yourdon, advocate; yet the belief in this way of doing things is so strongly ingrained in our thinking that it pervades both the way we ask questions and the questions we ask. We believe that these rationalistic traditions have led us away from expanding our horizons and asking new questions about the design process. We don't expect the reader, or ourselves, to suspend disbelief and simply discard current system methodology, but our intent, rather, is to show that by looking at the assumptions behind this rationalistic tradition we can better understand how we get trapped and limited in our thinking—limits that may appear as "mistakes" in our practice, but are, in fact, embedded parts of the rationalistic world view and the accompanying system approach. As Pelle Ehn and Morten

Kyng (1984) point out, many systems that are designed to reduce the need for skilled and experienced workers do so because the designs are based on this rationalistic world view which tends to make designers:

- view the application from the top of the organization,
- view the organization as a structure, whose important aspects may—and should—be formally described,
- reduce the jobs of the workers to algorithmic procedures, and thus
- view men and computers as information processing systems, on which the described data-processing has to be distributed. (p. 215)

Russell Ackoff, a well-known proponent of the systems approach, was well aware of the problems embedded in the dominating world view in the field. Trying to get system designers to focus on this issue, he said:

> We fail more often because we solve the wrong problem than because we get the wrong solution to the right problem....The problems we select for solution and the way we formulate them depends more on our philosophy and world view than on our science and technology. (p. 8)

This focus on world view reminds us that the issue of selecting problems within the workplace is heavily laden with cultural, political, and economic values. As computer system designers we can no more jettison our past than we can ignore the traditions of computer system users. Our approach in this book is to try to realistically appraise the way we work with a clearer understanding of the politics of the workplace and a careful eye on the assumptions we bring with us. The case studies in the following chapters are, like most workplace situations, potentially explosive in terms of contrasting management's objectives for the system with users' skills and interests. We won't attempt to sidestep this issue, but as we mentioned earlier, we take seriously the idea that systems should be concerned with the quality of work, not just its quantitative output. Thus we are asking different questions and perhaps looking for "solutions" to other problems.

In this book we contrast traditional system development with our own evolving cooperative or participatory approach. We know that we can't make a clean break with the Cartesian dualism that has dominated rationalistic thinking in the past. As Kuhn (1970) highlights in *The Structure of Scientific Revolutions,* paradigm shifts evolve through contradiction over time. But we will certainly try to highlight ways in which our approach differs from the rationalistic world view and to point to emerging contradictions. In doing this we pay particular attention to the complex social relations of the workplace, and the need to use techniques that support involvement, rather than the detached reflection of the Cartesian scientist.

How Did We Get Here?

Like most projects, the writing of this book was in part serendipitous. Some of what we learned we learned through experience, and, as the chapters indicate, not all of it can be counted among the success stories. Some of our learning came through groping our way through the literature, as we attempted to find theories and examples that made sense of our experiences. Here, we present some of the strands that came from our earlier work and our reading, in order to provide an introduction to ideas presented in the book.

An obvious starting point for many authors in this book was in trade union-oriented projects in Scandinavia. Employee influence through unions and collaboration with management is a well-known part of the practice of social democracy in the Scandinavian countries. In the early 1970s, when new legislation increased the possibilities for worker influence, this strategy, called co-determination, was supplemented with a series of projects set up by central and local unions independent of employer organizations and management. In these projects, workers aided by consultants and researchers struggled to develop a better, more coherent platform for worker influence on the use of new technology in the workplace. The need for a new platform, based on the workers' own perspective, was described by Kristen Nygaard (1979) in the following way:

> It became clear that (the new possibilities for) industrial democracy would only give useful results if .. new knowledge (was) built up and a broad activity base established among the members of the trade unions.... Also, it was felt that workers would risk brainwashing themselves if they tried to assimilate the existing knowledge about the effects of (computer) systems as a starting platform. (p. 95)

New work practices, focusing on group work and the development of local resources for action, were being shaped, tried out in practice, and reshaped in the projects. Some of the work groups produced criteria for better working environments and suggestions for systems to support groups of workers planning their own work. As a result of the first of these projects, the existing legislation on worker influence was supplemented by Technology Agreements that gave workers a direct say in the development and use of technology in their workplaces. This also led to an extensive series of union education programs (Ehn & Kyng, 1987).

In the 1970s these early projects introduced the notion of worker participation in decisions about technology, but they ran into a series of problems. Writing about some of these projects, Pelle Ehn and Morten Kyng in *Computers and Democracy* (1987) noted:

> From a union perspective, important aspects like opportunity to further develop skill and increase influence on work organization were limited. Societal constraints, especially concerning power and resources, had been *underestimated*, and in addition the existing technology constituted significant limits to the feasibility of finding alternative local solutions which were desirable from a trade union perspective. (p. 29, italics added)

In short, whereas workers had a legal say in workplace technology, the laws did little to shift the balance of power from a managerial perspective. And, as discussed in the last section, the rationalistic tradition embedded in computer system development did little to give workers a voice in putting forth their own ideas when trying to agree on the introduction of new technology. As in the U.S., this was reflected in the tools of system development, which emphasized developing technical specifications rather than seeing the system from the perspective of the users.

By the early 1980s, a "second generation" of design projects was initiated in Scandinavia. These projects focused on using skill as a wedge to push computer system design more towards a users' perspective. They had as their theoretical starting point ideas put forth in the book *Labor and Monopoly Capital* by Harry Braverman (1974), which documented how capitalism increasingly took skill away from workers and brought it more and more within the hands of management. This process, called deskilling, intensified division of labor and resulted in work processes that were extensively routinized. Indeed, in the computer field in the 1960s and 1970s, this process of routinization resulted in the kinds of rationalized, cut-and-dried systems with which mainframe systems became synonymous (Greenbaum, 1979); systems that separated, for example, data entry from data verification, or customer relations from record keeping.

Braverman's point was that the act of dividing labor and deskilling workers was dehumanizing. The second generation of Scandinavian design projects took the issue of dehumanization and put it on the table as a central problem in the design and use of computer systems. Thus, to put some muscle on the bones of the Technology Agreements, the issues of *quality of work* and *worker skill* were put into the foreground of the computer system design projects. An example of this was the UTOPIA project, named both for its ideals and as an acronym for its use (Bødker, Ehn, Kammersgaard, Kyng, & Sundblad, 1987; see also Chapters 7 and 9). In this project computer system developers and researchers worked with a group of typographers to help them formulate the ways that computer technology could be used to enhance their skill and better the typographic quality of newspapers.

In their quest to make workplace skill a more central aspect of computer system design, these project organizers ran into severe difficulties in trying to apply the tools and techniques of traditional system development. They experienced how limited these tools and techniques were in their ability to actually allow workers, as users, to express their ideas about skill. In particular, they pointed out how the notion of tacit skill—those essential, yet not easily articulated qualities that we use in daily work—was difficult for computer system designers to grasp using formal system specifications.

From the introduction of Technology Agreements in the 1970s, through the focus on skill in the early 1980s, computer system development in Scandinavia struggled with the elusive concept of user participation. In *Work-Oriented Design of Computer Artifacts* (1989), Pelle Ehn outlines the story of these changes and delves into some of the theoretical work that has helped some of the authors of this book reflect on their work practice and that of users. In general, these theories can be grouped under the philosophical heading of social construction, which sees our understanding of the world as generated by people (through their social interactions) rather than as a set of fixed, immutable facts (see Bruffe, 1986; Rorty, 1979). In contrast with the rationalistic tradition of computer science, social constructionist theory veers away from rigid poles like "objective-subjective," and steers toward understanding different, pluralistic perspectives of how we think and act. Seriously, system developers have little room to hide behind a mask of objectivity, for developers, like users, need to get involved in day to day activities and learn to share perspectives.

In order for this book to move from user participation to active cooperation between users and system designers, several more issues had to fall into place. Among these, in both the United States and Europe, was the influence of feminist thinking. In *Reflections on Gender and Science*, Evelyn Fox Keller (1985) points out that the traditions of science have historically been rooted in a language that places value on words like "objectivity," "reason," and "impersonal judgement." These terms are associated both with the way that "good" science is done, and with those characteristics that are usually associated with men. Thus, historically within our culture, to be emotional or personal has been considered "female," while rational and impersonal are more closely linked with male attributes. It is not surprising to find this gender-biased thinking within system development; indeed the rationalistic tradition of computer system development with its step-by-step impersonal and supposedly objective procedures is considered as "good" rigorous practice (Greenbaum, 1987). In searching for ways to overcome these biases some of the authors in this book have borrowed from activities that grew out of the women's movement in the 1970s, such as small group "consciousness raising" sessions that focused on getting women to "speak in their own voice," and encouraged participation by all (Bødker & Greenbaum, 1988).

Within the computer field, we were also helped by the thinking of Terry Winograd and Fernando Flores in their book *Understanding Computers and Cognition* (1985). The book, subtitled *A New Foundation for Design,* critiqued the rationalistic tradition and laid the theoretical groundwork for helping us understand that "in designing tools we are designing ways of being" (p. xi). Indeed, their emphasis on the importance of action and on the difficulty of articulating assumptions grounded our work in the importance of the action-based techniques in design. Hubert and Stuart Dreyfus in *Mind over Machine* (1986) added to our theoretical understanding that skill and expert performance cannot be captured as a set of formal rules. Their work also helped us to focus on how skills are performed within the context of specific situations.

In *User Centered System Design*, Donald Norman and Stephen Draper (1986) brought together a range of articles that address the issue of design from what they call pluralistic perspectives—looking at design with the user's point of view in mind. As one of the first major American books to place users in the foreground, they take Human-Computer Interaction (HCI) as a focal point for designing for people, not for technology. In Chapter 2 of this volume, Liam Bannon, also a contributor to that book, investigates some of the ways that we can now go beyond conventional HCI studies in order to move from user-centered design to more cooperative design.

In 1986 in Austin, Texas, the first biannual Conference on Computer Supported Cooperative Work was held (see CSCW Proceedings, 1986, as well as CSCW Proceedings, 1988, from the second conference). While the conference was not about cooperative design, it did bring Americans and Europeans together to talk about design of systems that support the cooperative and interactive nature of workplace activity. In 1989 the first European Conference on this issue was held in England (EC-CSCW Proceedings, 1989), and in 1990 the first Participatory Design Conference was organized in Seattle, Washington (PDC Proceedings, 1990). To us, it is clear that the time is ripe for mixing and matching the Scandinavian traditions discussed in this book with American projects that focus on user involvement. Some American system designers have said that although they liked the Scandinavian approach, they feared that it wasn't applicable in the United States because of the weak trade-union movement.[1] Scandinavia's high degree of union membership, however, may only be a partial blessing for participatory design, for in those countries, as in the U.S., established unions sometimes tend to be stuck in their ways. On the other hand, American discussions about cooperative work and team approaches to work tasks, while perhaps overstated in the business press, nevertheless point out some fertile ground for planting seeds for cooperative design.

What Have We Learned?

> About 20 years ago, a sign in the "in-room" of a large computer center in Denmark read: "The impossible takes an hour—miracles take two." In the output room another sign stated: "You are lucky to get anything at all!"

Traditionally, people in the computer field have been quite optimistic about what computers can do. To many users and managers, system developers, like salespeople, promise them everything but seem to deliver very little. Even during the 1980s, when system developers began to involve users in the design process, users continued to grumble about the "stupid mistakes" of the designers. Certainly the complaints made by designers about users are equally loud and numerous.

In this book we have no easy way out of this situation. As we found out, switching the focus from users as passive participants in systems design to active user-designer cooperation is not easy. We hope that the readers will not think that we are promising miracles. Yet from our actual project experiences, from our readings, and from our own theoretical work, we think that we have learned a good deal about what can be done. Here is a summary of the approach advocated in the following chapters. It grows out of our

---

[1] Obviously a number of other differences exist between design in the U.S. and in Scandinavia, differences which are important in relation to user/designer cooperation. A number of these, including the emphasis on "product-development" projects in the U.S. vs. the predominance of "in-house-development" projects in Scandinavia, are discussed in Grudin (1990) and Grønbæk, Grudin, Bødker, & Bannon (1990).

experience in applying the design ideals mentioned in the beginning of the chapter.

- The design process needs to start with an understanding of the use situation. Traditional system development advocates beginning with the identification of "the problem," yet problems out of context have little meaning. That is why we believe that examining the context and paying close attention to the situations in which computers will be used is an appropriate starting point. To do this effectively, designers and users will find themselves stumbling along some untravelled paths as they try to learn about each others' basic assumptions.

- When computer systems are introduced within an organization they change the organization. Likewise, computer systems are not static entities, but rather systems that adapt as they are used. This dynamic process of ongoing change means that as designers we need to better understand the organization, and design for ongoing change. In the past, changes in work organization have often been looked at as "unintended consequences" of a new computer system. We don't think so. By designing with an understanding of work practice and its organizational setting, we think that we can, at least, be less naive about future changes.

- The design process is firmly rooted in experience, not just rules. Most computer system methodologies recommend step-by-step procedures for carrying out the design and implementation of computer systems. We don't think that these rule-based approaches should be thrown out, but we do believe that relying more on the experiences of designers and users can lead us toward systems that are more suitable for the workers involved.

- Users are competent practitioners. With this in mind we need to design for their skill, knowledge, problems and fears. Rather than planning "idiot proof" systems, we think that skill and quality of work should be given priority. Of course, not all users (or designers for that matter) are equally skilled, but diversity of skill is among the things we can learn by more carefully reflecting on work practice.

Early in this century, the famous American pragmatist and educator John Dewey, made a number of comments that we feel can speak directly to design at work. In his critique of education, he warned that students were seen as passive receptacles into which static bits of knowledge were to be placed. For us as system designers we are sharply aware of the need for active learning during the development process in order to avoid the trap of seeing users as passive receptacles. In *Experience and Education*, Dewey (1938) argued that rote learning limits the "power of judgment and the capacity to act intelligently in new situations." (p. 27). Dewey, sometimes called one of the first social constructionists (Rorty, 1979), advocated making the learning process more active and firmly rooting it in the experiences of the teachers and students. As we reflect on the problems of computer system development today, Dewey's work makes lively reading. For example, he warns that "the central problem of an education based on experience is to select the kind of present experiences that live fruitfully and creatively in subsequent experiences" (p. 28). As the authors in this book set out to work with users in understanding each others' experiences, we are reminded that selecting and interpreting relevant experiences can be as difficult as selecting problems in the traditional approach.

For readers with a background in system development, we think it's worthwhile to present an overview of some of the major differences between our approach and the combination of approaches traditionally put forth in literature and practice today. Like all overviews it is rather simplistic, but we believe it can serve as a guide for understanding the book.

TRADITIONAL APPROACH *focus is on*	COOPERATIVE APPROACH *focus is on*
problems	situations and breakdowns
information flow	social relationships
tasks	knowledge
describable skills	tacit skills
expert rules	mutual competencies
individuals	group interaction
rule-based procedures	experience-based work

Figure 1. Contrasting Approaches.

Reframing from Within and Without

This book attempts to reframe our thinking about design at work with experiences from both inside and outside of the computer field. This process of reframing, or seeing things in new ways, is useful for presenting our ideas about experiencing the present and envisioning the future. Part I, Reflecting on Work Practice, is largely based on ideas from the social sciences and humanities and their contributions toward helping us understand current work practices, especially the use of language and artifacts and their relationship to design. Part II, Designing for Work Practice, takes a look at ongoing computer system design projects and focuses on ways that designers and users can work together to envision future use situations as well as adapt systems in on-going use. The browsing readers can pick and choose the chapters that apply to their interests, but we suggest that Chapter 7 be read as background to Part II, before browsing in that section.

In the first two chapters of Part I the authors present a broad, conceptual overview of ideas that influence how we look at work and the use of computers.

Liam Bannon, with a background in computer science and cognitive psychology, builds a bridge between more conventional Human-Computer Interaction studies and cooperative design. His essay travels down a road, which, as he points out, may lead traditional research in HCI to address the ideas of cooperative design developed in this book. Specifically, he is critical of traditions in both psychology and human factors that place too much reliance on controlled laboratory studies and not enough on the actual setting of work.

Eleanor Wynn, a linguistic anthropologist, focuses on the power of perspective and the involvement that anthropologists use in their field work. Her work, applied to an understanding of conversation among clerical workers, brings to the foreground the complexity and problem-solving capacity found in daily work. Using language as a handle on assumptions, she points to ways that we can better understand what takes place at work.

The next three chapters present research from projects in anthropology, linguistics, and organizational behavior, suggesting strategies that we could apply to better understand how people work, talk, and use artifacts, this includes hints on how such an understanding may be used in the design of computer systems.

Lucy Suchman and Randall Trigg tell us about a project which uses videotapes to help researchers reflect on work practices. Lucy Suchman, an anthropologist, and Randall Trigg, a computer scientist, give us suggestions for how interaction analysis in general, and videotaping in particular, can be used in developing an understanding of the situated use of artifacts. They also discuss how their techniques may be developed and applied in a participatory approach to design.

In Chapter 5, Peter Bøgh Andersen and Berit Holmqvist, both linguists, take us on a tour of their field, explaining concepts and methods that could be used as a basis for system design. They analyze work language in specific cases to help us understand how perspectives differ within organizational roles and with the task at hand. They continue by showing how these differences give rise to different demands which a computer system will have to fulfil in order to suit the work.

The last chapter in this part of the book takes a look at workplace cultures and lets us see how artifacts at work can be clues for understanding basic assumptions within the organization. Jesper Strandgaard Pedersen, with a background in organizational behavior, and Keld Bødker, educated in computer science, team up to give us examples of how shared values and beliefs can be looked at in small working environments. Their approach is different from traditional studies of organizational culture, which look at the values of a whole organization. By zeroing in on workplace cultures *within* organizations they give us ideas about how computers, as artifacts, could better reflect workplace values.

Part II opens with a chapter entitled "Setting the Stage for Design as Action" by Susanne Bødker and Morten Kyng, both computer scientists, and Joan Greenbaum, who has a background in system design and economics. This chapter gives an overview of the ideas and theories developed and used in Part II, and also looks at some of the common questions that people ask when trying cooperative design. The theoretical perspectives presented here build on the social constructionist views we outlined earlier. It begins with the notion that we only understand what we have already understood. One of the constant problems in system design, of course, is the issue of how users can imagine the future, and how designers and users can transform this imagination into practical computer systems. "Setting the Stage" gives some clues for what to expect when embarking on this process.

The subsequent chapters in Part II are laid out in a pattern which resembles the process that designers and users go through in designing, using, and modifying a computer system.

Chapter 8, by Finn Kensing and Kim Halskov Madsen, both computer scientists, suggests the use of Future Workshops and metaphorical design in order to help users generate visions about their organization and future computer use that transcend their current work practices. The Future Workshop technique uses specific methods to help people brainstorm about their current practices and its shortcomings and to find possibilities for future alternatives.

Chapter 9, by Pelle Ehn, educated in both sociology and computer science, and Morten Kyng, tells the story of how mock-ups were used in the early stages of design to help typographers envision the kind of hardware and software support they would need to enhance the quality of their work as well as of their products. The chapter gives examples of the ways that mock-ups and and even computers can be used to experience and envision different future possibilities before large amounts of money are invested.

In Chapter 10, Susanne Bødker and Kaj Grønbæk, also educated in computer science, present ideas and examples for developing prototypes in cooperation with users. Prototyping as a tool for trying out future systems has been gaining popularity. The authors' focus, however, goes beyond presenting prototypes as demonstrations *to* users, suggesting instead ways that users and designers can change prototypes cooperatively.

The next chapter is not on implementation, as one might expect in a traditional book on system development. The gap between the tools and techniques of implementation and those used in the different application areas is still so big that we have not yet attempted to bridge it together with the users, although current work on object-oriented programming may improve this situation. Therefore, in Chapter 11 we turn our attention to the process which, in recent years has become known as "tailoring," or adapting and modifying systems once they are in use. Austin Henderson, educated in mathematics and computer science, and Morten Kyng present arguments for the importance of continuing design in use. They go on to give examples of how users are to take an active role in continuing design in use.

The last chapter in Part II, by Pelle Ehn and Dan Sjögren, educated in economics, takes us back to the problems of developing new work practices when technology changes. Supplementing the preceding chapter, this chapter focuses on how users can enhance their work practices as new systems are adopted.

These chapters in Part II look at design from the perspective of users and designers working together. Although the perspective of managers and organizations is an important one, much has already been written about it. Our approach is to flip the pyramid and look at the problems from what is traditionally viewed as the bottom or the end user perspective.

The epilogue, Chapter 13, introduces many of the ideas we would have liked to put into the book had we had more time and experience. It tells the tale of a Future Workshop in which all the authors participated, in order to "envision" the book that we would like to have written. Entitled "Design by Doing," it takes us from our current experiences in cooperative design to an action-based approach that we would like to imagine in widespread use in the near future. It summarizes the cooperative approach discussed here, but it also presents our frustrations about the present and our hopes for the future.

As a note on terminology to the people from different backgrounds who will be reading this book, we mainly look at the process of developing multi-user computer systems in workplaces and organizations. We are writing primarily for the designers, researchers, students, and curious users who are involved in that dynamic intersection where technical support and actual use situations meet.

When we use the term computer system, we use it in the broadest possible sense. Just as a building is more than the sum of the wood, nuts, bolts, and plans used in its construction, a computer system is both the entity itself and the way it is used. It includes the use of the hardware, such as processors, printers, monitors, and keyboards; it encompasses the design of software, whether it is custom designed programs for large-scale projects or the selection of off-the-shelf packages; and, of course, it focuses on the people who are going to use the new system.

The specialists who design and plan the construction of a building are conveniently called architects, yet the legions of technical and analytical specialists who put together computer systems go by many names. Here, we refer to the technical people as *designers* in order to keep our eyes on the *process* of developing computer systems. In traditional terminology these system designers may be programmer/analysts, system analysts, system engineers, or consultants, and their tasks may range from conducting feasibility studies through design and implementation of systems. But as all system analysts know, regardless of their titles or range of experience, the process of taking a system from a set of vague ideas to the actual use of computer tools is not a straightforward, linear one. Noting this, we use the terms *design* and *development* interchangeably, because both are on-going creative activities.

Our use of the term *user* may create less immediate confusion. However, among the authors, our inability to find a suitable substitute for the word created quite a lot of frustration, because this word groups all the different kinds of competent practitioners together under one label. Thus, when talking about "designers and users," the specific words do not emphasize the participation of specific workers who, as users, are active in a design process. This being said, we hope that the message gets through despite the shortcomings in the language.

In many instances throughout the book, the words "computer system" could be replaced with words covering other kinds of technology. For the reader interested in a broad understanding of workplace and technology issues, we offer numerous examples that could be creatively applied to different kinds of workplace issues. Additionally, although most of the examples deal with developing custom-designed systems for networked workstations or personal computers, the knowledge gained from these examples is, we believe, applicable to situations where off-the-shelf software is developed or adapted.

Each of the chapters in the book applies some ideas about cooperative design *depending on the situations* in which the designers and users find themselves. This emphasis on situated design means that we can offer the reader no straightforward method or universally applicable set of tools. We invite you to read, reflect, and try out the ideas presented here, based on the workplace situations in which you find yourself. We hope that our experiences spark your curiosity about cooperative design, leading you to the suggested readings and to try your own applications.

References

Ackoff, R. (1974). *Redesigning the future. A system approach to societal programs*. New York: Wiley.

Bjerknes, G. & Bratteteig, T. (1984). The application perspective— Another way of conceiving systems development and edp-based systems. In M. Sääksjärvi (Ed.), *Proceedings from the Seventh Scandinavian Research Seminar on Systemeering, Part II* (pp. 204-225). Helsinki, Finland: Helsinki School of Economics.

Bjerknes, G., Ehn, P., & Kyng, M. (Eds.). (1987). *Computers and democracy—a Scandinavian challenge*. Aldershot, UK: Avebury.

Boguslaw, R. (1965). *The new utopians—A study of system design and social change*. Englewood Cliffs, NJ: Prentice-Hall.

Braverman, H. (1974). *Labor and monopoly capital—The degradation of work in the twentieth century.* New York: Monthly Review Press.

Bruffe, K. A. (1986). Social construction, language and the authority of knowledge. In *Journal of College English, 48,* 773-790.

Bødker, S. & Greenbaum, J. (1988). A feeling for system development work—Design of the ROSA project. In K. Tijden, M. Jennings, & U. Wagner (Eds.), *Women, work and computerization: Forming new alliances, Proceedings of the IFIP TC 9/WG 9.1.* Amsterdam: North-Holland.

Bødker, S., Ehn, P., Kammersgaard, J., Kyng, M., & Sundblad, Y. (1987). A utopian experience. In G. Bjerknes, P. Ehn, & M. Kyng (Eds.), *Computers and democracy—a Scandinavian challenge* (pp. 251-278). Aldershot, UK: Avebury.

CSCW (1986). *Proceedings of the Conference on Computer-Supported Cooperative Work, Dec. 3-5, 1986.* Austin, Texas. New York: ACM.

CSCW (1988). *Proceedings of the Conference on Computer-Supported Cooperative Work, Sept. 26-28, 1988.* Portland, Oregon. New York: ACM.

Daressa, L. (Producer). (1986). Computers in context. [Film]. San Francisco: California Newsreel.

DeMarco, T. (1978). *Structured analysis and system specification,* Englewood Cliffs, NJ: Prentice-Hall.

Dewey, J. (1963). *Experience and education.* New York: MacMillan. (Original work published 1938.)

Dreyfus, H. & Dreyfus, S. (1986). *Mind over machine—the power of human intuition and expertise in the era of the computer.* Glasgow: Basil Blackwell.

EC-CSCW (1989). *Proceedings of the first European Conference on Computer-Supported Cooperative Work, Sept. 13-15, 1989.* London. New York: ACM.

Ehn, P. (1989). *Work-oriented design of computer artifacts.* Hillsdale, NJ: Lawrence Erlbaum Associates.

Ehn, P. & Kyng, M. (1984). A tool perspective on design of interactive computer support for skilled workers. In M. Sääksjärvi (Ed.), *Proceedings from the Seventh Scandinavian Research*

Seminar on Systemeering, Part I (pp. 211-242). Helsinki, Finland: Helsinki School of Economics.

Ehn, P. & Kyng, M. (1987). The collective resource approach to system design. In G. Bjerknes, P. Ehn, & M. Kyng (Eds.), *Computers and democracy—a Scandinavian challenge* (pp. 17-57). Aldershot, UK: Avebury.

Greenbaum, J. (1979). *In the name of efficiency—Management theory and shopfloor practice in data processing work.* Philadelphia: Temple University Press.

Greenbaum, J. (1990). The head and the heart. In *Computers and Society, 20* (2), 9-16.

Grudin, J. (in press). *The development of interactive systems: Bridging the gaps between developers and users.* IEEE Computer.

Grønbæk, K., Grudin, J., Bødker, S., & Bannon, L. (forthcoming). Improving conditions for cooperative design—shifting from a product to a process focus. In D. Schuler & A. Namioka (Eds.), *Participatory Design.* Hillsdale, NJ: Lawrence Erlbaum Associates.

Hoos, I. (1961). *Automation and the office.* Washington, DC: Public Affairs Press.

Jackson, M. (1983). *System development.* Englewood Cliffs, NJ: Prentice-Hall.

Keller, E. F. (1985). *Reflections on Gender and Science.* New Haven, CT: Yale Univ. Press.

Kuhn, T. S. (1970). *The structure of scientific revolutions.* (2nd ed.) Chicago: University of Chicago Press.

Norman, D. A. & Draper, S. W. (Eds.). (1986). *User centered system design—New perspectives on human computer interaction.* Hillsdale, NJ: Lawrence Erlbaum Associates.

Nygaard, K. & Bergo, O. T. (1975). The trade unions—New users of research. *Personnel Review, 4* (2), 5-10.

PDC'90 Proceedings (1990). *Proceedings, Participatory design conference.* April, 1990, Seattle, Washington. Palo Alto, CA: Computer Professionals for Social Responsibility.

Rorty, R. (1979). *Philosophy and the mirror of nature.* Princeton: Princeton University Press.

Suchman, L. A. (1987). *Plans and situated actions—The problem of human-machine communication.* New York: Cambridge University Press.

Weizenbaum, J. (1976). *Computer power and human reason.* New York: W. H. Freeman.

Winograd, T. & Flores, F. (1986). *Understanding computers and cognition.* Norwood, NJ: Ablex.

Wittgenstein, L. (1953, 1963). *Philosophical investigations.* Oxford, UK: Oxford University Press.

Yourdon, E. (1986). *Managing the structured techniques.* New York: Yourdon Press.

Part I

Reflecting on Work Practice

2

From Human Factors to Human Actors: The Role of Psychology and Human-Computer Interaction Studies in System Design

Liam J. Bannon

Man is one of the best general-purpose computers available and if one designs for man as a moron, one ends up with a system that requires a genius to maintain it. Thus we are not suggesting that we take man out of the system, but we are suggesting that he be properly employed in terms of both his abilities and limitations.
E. Llewellyn Thomas, 1965

I believe that there needs to be a better understanding among researchers, and among many system designers too, about the "users" of computer systems and the settings in which they work. Part of the problem resides in an implicit view of ordinary people which, if surfaced, would seem to treat people as, at worst, idiots who must be shielded from the machine, or as, at best, simply sets of elementary processes or "factors" that can be studied in isolation in the laboratory. Although psychology, particularly as represented by the field of human factors (HF) or ergonomics, has had a long tradition of contributing to computer systems design and implementation, it has often neglected vitally important issues such as the underlying values of the people involved and their motivation in the work setting.

Understanding people as "actors" in situations, with a set of skills and shared practices based on work experience with others, requires us to seek new ways of understanding the relationship between people, technology, work requirements, and organizational constraints in work settings. Studying such a multi-faceted and multi-layered issue requires that we go beyond more traditional controlled laboratory studies that are the hallmark of experimental human-computer interaction (HCI) studies. In this chapter I recount some experiences, give some background on the field of HCI, and point out some problems. I suggest some alternative perspectives and directions for more fruitful research on, or rather with, people in work settings that may assist in the design of more usable and useful computer systems.

A True Story

I once worked in the human factors group of a large organization and was asked to build an interface to a particular system. I was given the manuals from the previous system, some reports by investigators of problems noted with this system, and some specifications of the new hardware and software capabilities and the new functional specifications. Fine, as far as it went. My wish was to go out and talk with users of the current system, to get a better "feel" of the situation they were working in, and subsequently to interact with them in an iterative design process in order to make a usable interface. To my astonishment, I was informed that not only could I not proceed in this fashion, but—ultimate irony—I was not allowed, for political organizational reasons, to meet with even a single user throughout this process! I complained, but being in a very junior position, I was overruled, so I then spent some months on what seemed to me an insane task: developing a logically coherent, though naive, mapping of tasks onto a menu-based interface based solely on the paper specifications I had been given. A report was duly completed and forwarded to another level in the organization, and much of it was actually implemented, to my amazement. Given my deskbound, limited knowledge of the actual use situation, I am sure that many of my "logically" elegant design solutions turned out to be hopelessly complex and/or inappropriate in the real work situation.

I do not want to claim that such a situation is the norm, but I do claim that this true scenario is not nearly as far from normality as one would hope. Granted, the fault here does not lie solely at the feet of the HCI community. Within this organization, however, that group must take some of the blame for not stressing an iterative design process that insisted on *active* user participation. This is partly due to a rather limited perspective on what users can offer in the design process, compounded by organizational politics that sometimes makes contact between designers and users a more difficult process than one would suppose. Readers might keep this little story in mind as they read the rest of the text, as the link between it and some of the attitudes and practices of designers and researchers discussed here should emerge. In the next section I take a short tour of the landscape of human factors and user interface design, note a few pertinent examples of terminology which betray a certain limited view of the people that we aim to design for, and suggest some alternatives.

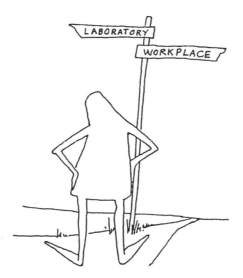

A Question of Perspective(s)

Language does not simply symbolize a situation or object which is already there in advance—it makes possible the existence or appearance of that situation or object, for it is part of the mechanism whereby that situation or object is created. (Mead, 1934, p. 78)

Replacing Human Factors with Human Actors

The terms used in a discipline often give a clue as to how members "see" that field. Once certain distinctions are accepted and become part of the standard vocabulary, these terms can become a barrier to the reality that lies outside. For this reason, it is a useful exercise to periodically reexamine the language used to express our understanding of the world. I have chosen to use the terms human factors and human actors in the title of the chapter as I believe it highlights a difference in the perception of the person; the former connoting a passive, fragmented, depersonalized, unmotivated individual, the latter connoting an active, controlling one. I claim that traditional human factors work, although it undoubtedly has merit and has produced many improvements to existing technological systems, is often limited in scope with respect to its view of the person. Within the HF approach, the human is often reduced to being another system component with certain characteristics, such as limited attention span, faulty memory, etc., that need to be factored into the design equation for the overall human-machine system. This form of piecemeal analysis of the person as a set of components de-emphasizes important issues in work design. Individual motivation, membership in a community of workers, and the importance of the setting in determining human action are just a few of the issues that are neglected. People are more than a sum of parts; be they information-processing subsystems or physiological systems, they have a set of values, goals, and beliefs about life and work. This more encompassing interpretation of the human factors field is not signaled by the name, nor by the bulk of studies appearing under this topic in the journals so designated. By using the term *human actors* emphasis is placed on the person as an autonomous agent that has the capacity to regulate and coordinate his or her behavior, rather than being simply a passive element in a human-machine system. This change in terminology may help in adjusting one's perspective, emphasizing the holistic nature of the person acting in a setting, and contrasting with the view of the person as a set of information processing mechanisms which can be analyzed in the same manner as the information processing mechanisms of the technology.[1]

Re-thinking the Concept of "Users"

Another term ubiquitous in articles about the HCI field that deserves scrutiny is that of *users*. This general term refers to all people who use a particular computer system or application. It can be distinguished from the term "operators" in that the latter implies a greater involvement with the machine or system, presumably one where the person is more uniquely assigned to the device. The focus of the system design or HCI research group is biased towards the technology; the view of people is often simply as "users" of this piece of technology, and "naive users" at that. This can lead to problems. People may not know the technology, but they are not naive as to their work; rather it is the system designers who are "work naive." There is nothing inherently wrong in taking this stance—though in talking of naive users, I prefer the less judgmental terms "casual" or "discretionary users"—for periods when designing computer applications, but there is a danger in thinking of people as *nothing but* users. In fact, it is often still the case that computer users need to make some modifications to the system in various ways, tailoring the system before it is truly usable (see Chapter 11). So in a very real sense *users are designers* as well. Focusing on people simply as users can also blind us to the fact that

---

1 Similar concerns about the role of human factors, including reference to this issue of human actors, can be found in the work of Niels Bjørn-Andersen. See, for example, the paper "Are human factors human?" (Bjørn-Andersen, 1988).

the user's view of the technology we are developing may be very different from that of the designer's view. The user is often a worker who has a set of tasks to perform, and the use of the computer may be only one element necessary to the accomplishing of this work. We can become focused in design on one particular application and forget that the user has needs differing from that of simply performing the set of system operations. For example, Whiteside and Wixon (1987) note one case where a word processing clerk spent a lot of time manually counting the lines she had typed because she was paid by the line, but the particular system did not have a facility to do this (relatively straightforward) task. Often the user is involved in multi-tasking not just on the computer, but also with co-workers, perhaps with clients through the phone system, and sundry daily happenings. Neglecting this rich and complex set of "nonessential" but truly everyday work interactions can lead to unworkable or unwieldy systems (see Cypher, 1986). It is the ability to understand the user perspective, to be able to see a problem from other than the system viewpoint, to be able to empathize and work with the users of the intended system that marks the good design team.

Users are not Idiots

Although viewing computer users as naive is bad enough, viewing them as idiots is even worse. Apart from the values question that arises, there are clear design implications if this faulty view of users is implicitly adopted. Just because users do not understand how the machine works or have difficulty with the system designer's terminology, does not imply that they are stupid, as some developers apparently conclude, if we are to judge from the systems that are at times designed. The system design team should start out with the understanding that workers/users are competent practitioners, people with work tasks and relationships which need to be taken into account in the design of systems, and with whom they must collaborate in order to develop an appropriate computer system. The idea that we must design systems so that "any idiot can use them" bears close scrutiny. Taking this as a serious design goal can often result in systems that necessarily produce such stupid behavior (see Bannon, 1986a, for further comment). In addition, as noted in the opening quotation, a consequence of trying to make such a system is that an incredible amount of "intelligence" must go into its initial design and maintenance. Taken to the extreme, we have the prospect of artificially intelligent systems operated by morons—an absurd scenario. Fortunately, in recent years this particular problem seems to have diminished, probably as a result of designers and others finally developing a better understanding of the user's perspective. A slightly less obnoxious version of this can be seen in the overemphasis on "easy to learn" systems in HCI research. Certainly, there are some applications where a minimal interface that is easy to learn is desirable, but this is not usually the case when dealing with systems that will be used by people in their everyday long-term work. We need to pay attention to the capabilities of the system and allow users greater flexibility and expressiveness in their use of it.

Allow for Active Users

Whereas focusing attention on the user may be a positive step, users are not simply passive objects that others must study and design for, as some accounts would have it. People are, or can become, *active agents*. They often wish to accomplish tasks, to understand what is going on, and are willing to jump ahead and explore the computer system on their own if, for example, the tutorial material is unclear or pedantic. If the system does not give an explanation for its behavior, the user will often try to make one up in order to render the doings of the system comprehensible. People are always struggling to make sense of their world. Developing instruction sequences to be followed by rote, with inadequate explanation, fails to satisfy or even to work in most real situations. For example, in discussing LisaGuide, an on-line tutorial for the Lisa system, Carroll and Mazur (1986) note one user's reaction: "I'm getting impatient. I want to do something, not learn how to do everything." (p. 39). Understanding the needs of active users has been the subject of investigation by psychologists, with Carroll and his colleagues prominent among them (Carroll & Rosson, 1987). Here are the words of an office worker, a user whom I interviewed, which capture her wish to understand, and to learn what's going on.

> People have to know how to understand, they have to be able to rationalize things out, to work things out, in a logical sequence. If they don't understand something, or how it works, then they have to go back to either just learning it by rote, or by asking, or by just making the mistake, and going back, and asking, and then correcting. But if you understand the system a little bit then sometimes you can think it out ..... and if you can reason it out then it stays with you longer—it's easier to understand, to work with. But if you are just learning piecemeal, then you can't. If you are just learning by rote.... someone tells you, this is the way to do it, then you can memorize it, but you'll never fully understand it, and [never] be able to expand from there.

In summary, we can see on reflection that some of the concepts and terms common in the system design and HCI research fields contain at least an implicit, if not explicit, perspective on users of computer systems that I would claim is inaccurate and liable to lead to faulty design decisions. Likewise, within the theory and research areas of HCI, this perspective has led to a concentration on issues that can be studied in the laboratory at the expense of other concerns that might be more crucial in actual work settings. But exactly what is this HCI field, and how did it come about? In the next section, I give a brief account of the origins of the field, before proceeding to discuss the current status of HCI—some results, limitations, and possible future directions.

The Field of Human Factors and Human-Computer Interaction: Some Background

This section gives a short account of the changing relationship between people and the machines with which they work, and the contribution of human factors studies in these changes. It is evident that unless a process is totally automated there will still be a need for some human intervention. How to divide the work between the human and the machine becomes an important task. Traditionally, in human factors, this "allocation of functions" task proceeds according to sets of guidelines about general human and machine performance capacities. We also need to design an "interface" between the machine and the person—knobs, dials, displays, controls—so that they can interact at some level. In the early days of this century, the focus was on how to get a machine to perform to the required functionality. The human component was often reduced to yet another "cog" in the overall machine—one that was, at times, more expendable than the machine. The slow improvement in the conditions and nature of work for people operating with machines came about not simply for altruistic reasons, but partly also in the search for greater efficiency of operation. Poorly trained or motivated operators and poorly designed equipment could cause breakdowns in the smooth functioning of the industrial process. Better designed controls and tasks that reduced both mental and physical strain on the operator allowed for improved performance of the human-machine system. The attempt to fit the machine to the skills and limitations, both physical and mental, of the operator developed into a new field of applied study early this century. This new field was called *human factors engineering* in North America, and similar efforts in Europe became known as *ergonomics*, from the Greek words *ergon*, meaning work, and *nomos*, which can mean law or knowledge. People working in this field usually had a background in either behavioral science or industrial engineering. Physiologists and medical practitioners also contributed to the understanding of human capabilities and limitations in work settings—effects of stress, psycho-motor ability, perceptual acuity, mental processing workloads, etc. Out of this observational and occasionally experimental discipline arose a body of knowledge that could be useful in the design of complex human-machine systems (see Van Cott & Kinkade, 1972).

In the early days of this century the problem was to build machine systems to do something useful. The focus was on the machine performance, as I noted previously. Machines were not being substantially modified every year, so the *ease of learning* of the system was not a high priority. Training was a one-shot process. People could be trained to perform whatever operations were required and subsequently serve as operators of the machine. As computing developed, the separation between operating and programming developed. Focus was still on the functionality of the software rather than on its ease of use. Programmers spent years learning arcane languages to communicate with the system. The user community started to change from one that was focused either on studying the properties of the machines themselves (computer scientists) or on application programmers that mediated between user needs and the computer system. A growing number of people using computers could be classified as *discretionary* users, people who saw themselves as having a job or profession that was not primarily geared to the computing medium itself, but who used it directly as a tool in their everyday work. However, these people were frustrated by the difficulty of learning how to program computers for their work. With the advent of the personal computer over a decade ago it was obvious that their widespread acceptance would become more dependent on their ease of learning and use. Not everyone was willing to learn something akin to IBM's notorious Job Control Language (JCL) in order to operate a personal computer.[2]

The field of human-computer interaction (HCI) emerged in the early eighties partly as a response to these changing conditions. It was linked to, but somewhat distinct from, its human factors pro-

---

[2] Paradoxically, this emphasis on making systems easy to learn and use has now resulted in an over-emphasis in research and design on this aspect of computer use, with a consequent decline in attention to how to allow for the growth of competence and skill on many computer systems. I elaborate on this point later in the chapter.

genitor in both its eclectic makeup and its more theoretical bias.[3] The older field, somewhat dismissively referred to as "knobs and dials" psychology, was seen as lacking in theoretical motivation by cognitive scientists. What was required was a better *cognitive coupling* between the human and the new universal machine, the computer, and not simply better designed surface characteristics of displays. Software engineers were also involved, as they were experimenting with the design of highly interactive interfaces and were concerned about how to conduct dialogues with users and present complex information effectively to them on graphic displays. Over the last decade the area of human-computer interaction has grown enormously, both within academic research environments and corporate research laboratories. This commercial concern was a major impetus for HCI studies, and "ease-of-use" and "user-friendliness" have become advertisements for particular computer systems.

Beyond Current Conceptions of HCI

Despite the legitimate advances that have been made in various arenas of human-computer interaction (see Shneiderman, 1987, for a collation of some of this material), there has been serious criticism of the field for its lack of relevance to practitioners in system design. Despite the widespread interest, no clear set of principles has emerged from this work. The experience of certain designers has been loosely codified, various long lists of design guide-lines are available, and a large number of evaluations of existing systems have been produced; but the attempt to place this applied science on a more rigorous footing has been difficult. Gray and Atwood (1988), in a review of a recent collection of papers (Carroll, 1987), noted the lack of any examples of developed systems in the papers and the general lack of contact of the work with "real world" design situations. They explicitly state that the skeptical designer will not be convinced of the relevance of the cognitive sciences to the design of better human-computer systems based on this work. Within this collection, a discussion section by Whiteside and Wixon (1987) makes a number of pointed observations about the limitations of cognitive theory in its application to everyday design situations.

If there is not too much in the way of theory that is directly relevant, can we utilize some of the sophisticated methods and

---

[3] Earlier human factors "man-machine communication" (sic) groups, were mainly working within military high-performance environments e.g. fighter aircraft cockpit displays. Another traditional area of concern for ergonomists was in general occupational heath and safety studies, e.g. determining noise safety levels in factory settings.

techniques used in psychology to analyze user behavior, so that designers could find out how users perform on different versions of a system, or different prototypes, and not have to rely on intuition? Standard lab experimentation is too limited, costly, and time-consuming. What is needed are quick and dirty methods that can give rapid feedback to designers about the utility and usability of their products. This is happening in practice currently, in the work on *usability*. Design involves making many tradeoffs, and in many instances empirical data can assist in determining the appropriate choices. Such tasks as setting up small in workplace empirical studies, noting reactions, user preferences, taking verbal protocols from subjects to see how they view the application, etc., are all investigative methods familiar to psychologists which can be useful in an iterative design process.

Cognitive psychologists have also studied particular issues; for example, we now know a lot about how people learn to use word processors, about the kinds of errors they make on different systems, about the mental models they attempt to construct of systems; but the application of such findings to new situations is not always obvious. From the perspective of the designer, the work to date can highlight some pertinent issues, but we are still impoverished in our search for new ways of thinking about and developing systems. The hoped-for contribution of HCI to the design of completely novel interfaces has not yet materialized, though some believe that it may come, given a shift in priorities among psychology researchers (Carroll, 1989).

I believe that there are a number of limitations in much of the current work on HCI that needs to be remedied for the field to be more useful to designers in practical situations. In the following paragraphs I suggest some changes in direction within the field of HCI that might help bridge the gap between theory, experiment, system design, and the actual work setting.

From Product to Process in Research and Design

By this I mean that more attention needs to be paid to the *process* of design, that is, working with users in all stages of design, to see the iterative nature of design and the changing conception of what one is designing as a result of the process itself. This is in contrast to a view of design that proceeds from a set of fixed requirements without iteration and without involvement of the users. This change in orientation has been evident for some time in much of the system design work within the Scandinavian tradition (see Chapter 9). It is also evident in the work of Jones (1988) and Floyd (1987), from industrial design and software engineering backgrounds, respectively, and by Bannon and Bødker (1989) in the field of HCI.

From Individuals to Groups

The majority of HCI studies to date take as their focus the individual user working on a computer system. This research focus totally neglects the importance of coordination and cooperation between work processes that is necessary in many work situations. Indeed, here again the applied field has been more astute than the theoretical. System designers have been aware of this coordinated aspect of work activity and have tried to support it from early on, albeit rather crudely. For example, much of the office automation work in the 1970s attempted to model work flows, but in too rigid a manner, thereby not allowing the human "components" in the overall system enough flexibility to make the system work. With a better understanding of how work gets accomplished coming from the research of Suchman and Wynn (1984), designers have a better "model" from which to build. The word "model" is in quotes here because such research shows that a strict model of human action in most work situations is not possible or appropriate; rather, human action is driven by the concrete situation that exists at any moment and is constantly changing. This implies that we should *support office workers in their activities*, rather than building *office automation systems*. Extending the focus of concern from the human-computer dyad to larger groups of people and machines engaged in collaborative tasks is an important area for research in the next period. The quick growth of this field, labelled Computer Support for Cooperative Work (CSCW), attests to its importance.[4]

From the Laboratory to the Workplace

Much of the early research done in the HCI field was confined to rather small, controlled experiments, with the presumption that the findings could be generalized to other settings. Examples of such studies were those done on command naming conventions (see Barnard & Grudin, 1988 for a review of this research). It has become increasingly apparent that such studies suffer from a variety of problems that limit their usefulness in any practical setting. First, by the time these studies are done the technology often makes the original concerns outdated. Witness the disappearance of line-oriented editors from commercial systems just as a "scientific" understanding of how people used them was being developed. Newell and Card (1985) refer to this problem as the race between the tortoise of cumulative science and the hare of intuitive design. Important contextual cues for the accomplishment of tasks were often omitted in this

---

4 See Greif, 1988, for a selection of papers in this area, and Bannon & Schmidt (1989) for an overview of the field.

transfer from the real world to the laboratory, so the results of the lab studies became difficult to apply elsewhere. Increasingly, attention is shifting to in situ studies, in an effort to "hold in" the complexity of the real world situations, and a variety of observational techniques, especially video, are being employed to capture activities. We can see an increasing focus on the concept of "usability" among the research community—whether people can and do actually use the resulting systems designed for them.[5] From a design perspective, this means that we need a prototype or test system for users to experience in order to get information on the usability of the resulting system. There are many examples from recent large development projects that point out the importance of empirical methods for getting feedback on design decisions from users. For example, on the path-breaking Xerox Star system, both formal experiments (Bewley, Roberts, Schroit, & Verplank, 1983) and informal studies were used extensively. On the Apple Lisa, the designers note "Another thing we have done is user tests—taking our ideas and bringing in naive users and sitting them down and seeing what their impressions are. This has caused some changes, and I think that's all shown in the quality." (BYTE Interview, 1983, p. 100). Having psychologists on the design team can thus assist in the planning, conduct, and evaluation of such studies, whether or not detailed manipulations and data analyses are planned.

From Novices to Experts

The majority of experimental studies in HCI focus on first time learners of computer systems or applications. Typically, performance is monitored for the first hour or two on the system. Exceptionally, perhaps, use of an application is observed over a few days, but rarely for periods longer, such as weeks, never mind years. This has been due partly to the ease of obtaining naive subjects for experiments from subject-pools at universities and employment centers. While granting that there is some need for studying such users, the paucity of studies that are concerned with the process of development of expertise with a computer application is remarkable. The issue is not simply that expert performance needs further examination, but that we need to pay greater attention to how users become skilled in the use of the computer application.

---

[5] For an excellent tutorial on designing for usability see the chapter by Gould in Helander (1988). The chapter by Whiteside, Bennett, and Holtzblatt on usability engineering in the same volume is also worth studying. What these articles note is how difficult it is to evaluate the usability of an artifact without investigating the situations of use of that artifact. See Bannon & Bødker (1989) for an extension of this argument.

What obstacles or incentives are there within the system to encourage the growth of competence? Additional issues relate to the difference in system learning and use between freshman university students and particular work groups with their own already-established set of work practices which may hinder or support learning and development of competence on the computer system.[6]

From Analysis to Design

Early human factors work tended to focus on evaluation of existing systems, and analysis of features that had been found in the use situation to be good or bad from the point of view of the user. However, the concern of people in HCI now is how to build better artifacts. We don't just want to know about systems after they have been built, we want to know how we should build them in the first place, and even what we should build.[7] HCI should be a design science: "Design is where the action is," to quote a memorable phrase of Allen Newell. The question is how can HCI contribute to the design of more usable and more useful artifacts? Newell and Card (1985) argue for the importance of giving designers approximate calculational models of the person performing a task for use in making design decisions about the human-computer interface. Many groups are currently active in such user modelling, looking at the structure, content, and dynamics of individual user cognition at the interface. Much work in the area continues the GOMS (Goals, Operators, Methods, and Selection rules) model tradition of Card, Moran, and Newell (1983), extending it in various ways. The practicality and utility of such low-level calculational models in actual design has been the subject of some debate (Carroll & Campbell, 1986). While some reject this approach as narrow and time-consuming, others have pursued even more generalized user models as design support aids. For example, the concept of Programmable User Models (PUMs) (Young, Green, & Simon, 1989) is based on a generalized architecture of human cognition. This approach to assisting system designers in understanding users' needs appears to be unduly narrow. Rather than moving designers closer to actual users, such a device, if it existed, would seem to support the view that actual contact with real users was unnecessary. The designer could just program the PUM in order to understand the constraints

---

[6] For more on this issue of qualifications and computer use, see Bødker (1989).

[7] Given how often new design emerges from old however, this distinction between analysis or evaluation of existing systems and design of new ones is perhaps too strict. In many cases, evaluation of current systems can provide important "feed-forward" into the next design as well, and usually does so in practice.

on usability, and potentially not have to observe actual users at all. The very vision of a PUM seems a rather abstract view of human activity in the world, and to imply a rather strange relationship, or lack of one, between designers and users. I would claim that both the GOMS work and that on PUMs support a human factor, rather than a human actor perspective discussed earlier. Such work can never replace prototyping and actual empirical user testing, although it might have a role at a certain stage in the design of a new system.

From User-Centered to User-Involved Design

As a first step towards focusing the attention of the designer on the needs of the user and away from a concentration on the hardware, there has been an increasing emphasis on taking a "user-centered" approach to design (Norman & Draper, 1986). Exactly what the term user-centered system design means, or how it can be achieved, is far from clear. In some cases it dissolves into platitudes such as "Know the User." Such kinds of general guidelines are of little use in practical situations of design due to their lack of specificity. Gould (1988) discusses how important it is to have an early and continuous focus on users, to develop iterative designs, and to have early and continuous user testing. This is a step in the right direction: Users are being given a larger role in the design process, but it is still a relatively passive role. Although actual participation by users on the design team is mentioned, it does not figure prominently in this user-centered approach. A more radical departure from current thinking within the mainstream HCI world is to look at users not simply as objects of study, but as active agents within the design process itself. This involvement of users in design is both a means for promoting democratization in the organizational change process and a way to ensure that the resulting computer system adequately meets the needs of the users. It is this approach that has been the hallmark of many of the Scandinavian studies mentioned in this book and evident in its chapters.

From User Requirements Specifications to Iterative Prototyping

Over the years, it has been acknowledged that the standard way of representing user requirements in the functional requirements specification document is often inadequate, and the question has begun to be asked whether this is because of some problems in the way of doing the studies locally, or whether there is a fundamental problem with the very assumption that we can map out in advance users' needs and requirements successfully through simple techniques of observation and interviewing. In Part II of this book, it is argued that users need to have the experience of being in the future use situation, or an approximation of it, in order to be able to comment on the advantages and disadvantages of the proposed system. Thus, some form of mock-up or prototype needs to be built in order to let users know what the future use situation might be. In the construction and evaluation of such prototypes, the skills of the psychologist can be useful. This particular issue is a part of the previously mentioned shift from a product to a process orientation in both research and development of systems.

Conclusion: HCI in System Design

Some distinctions can be drawn between different kinds of system development projects.[8] In the development of products the need to have a user-friendly interface is very high, as it can determine the success of the product in the marketplace. At the same time, the exact user group for the newly designed product is not clearly specified in advance. This can have implications for user-involvement in design, as discussed throughout this book. However, it does not invalidate the need to work with future users; it just means that we cannot be as sure about satisfying their needs as fully as we can on a system development project oriented towards a single application for a particular, specific group. We should be able to find representative users to at least allow us to iteratively prototype the design. Much of this book is dealing with another form of project, specific development projects, where the involvement of users is the least problematic because, at least in theory, we know who they are. In some projects, requirements specifications may be drawn up by third parties, other than the users and developers, and this can make it difficult though not impossible to utilize some of the user-involvement techniques outlined in Part II.

Understanding the needs and the tasks performed by the user is basic to the system development process. However, it is a mistake to think that simply having a human factors person on the design team is by itself sufficient to ensure that the human factor has been adequately taken into account, in the sense that is being discussed here. Even in companies where there exist groups specifically targeted to give human factors advice on projects, one often finds that they have little influence over the design process, and are often regarded as "add-ons" by the engineering staff. This state of affairs has sometimes been encouraged, unfortunately, by the human factors personnel themselves, who often seem unwilling to understand the complete project or product, but focus on the narrow aspects that

---

[8] These distinctions originate from remarks by Jonathan Grudin, and have been developed in subsequent discussions with Susanne Bødker, Kaj Grønbæk, and Jonathan Grudin.

are adjudged to require human factors input. Such an attitude can also be forced on human factors people as a result of the structure and functioning of the organization, with project support from the HF unit only allowed at certain periods, and the physical separation of the HF department from that of the software engineering group.

Human factors, or ergonomics considerations, are often incorporated into the design process simply as a set of specifications to which the delivered system must adhere. The actual work of the human factors personnel is seen as operator task analyses to be fed into these specifications, and perhaps some interface retouching near the end of the development cycle, when the system design has already been fixed. In general, the role of these people has been seen as ancillary to the main task of building the system. The role of HF or HCI in system design today should be more fluid and pragmatic. Input is vital in discussing the initial capabilities of the system and its required functionality, persisting in the development and evaluation of prototypes, and in final screen layout considerations.

What is being advocated here is an approach which, although acknowledging the contribution that different disciplines can make to the design process, ultimately depends upon the users themselves to articulate their requirements, along with the system design team composed of a variety of specialists acting in the capacity of consultants to the project. Design teams and users must be prepared to acknowledge each others' competencies and to realize that effort must be made by both parties to develop a mutually agreeable vocabulary of concepts that can be shared by the different groups comprising the project. It is no easy task for different disciplines and work activities to accomplish this; additional research would be valuable in this area.

Some of the efforts in Scandinavia on involving users in design (see, e.g., Bjerknes, Ehn, & Kyng, 1987) provide a promising start towards the alternative system design paradigm advocated here, at least for project system development. Within such an approach, the starting and end point of the design process is with the users themselves, from what they require, to how they evaluate the prototype, and the iterations that follow. Along the way, the services of a variety of disciplines may be required, not just those of the software engineer and the ergonomist, but also perhaps architects, sociologists, and anthropologists. These disciplines should come together in the overall design process as required, and not as dictated by some arbitrary flow model by which the system design gets handed around sequentially from one discipline to another. It is from the mutual interaction of these different perspectives, including that of the end users, focused on a particular design project, that good design may emanate.

Acknowledgments

Thanks to Niels Bjørn-Andersen, Susanne Bødker, Jonathan Grudin, Randy Trigg, and the Editors for helpful comments on earlier drafts.

References

Bannon, L. (1986a). Issues in design: Some notes. In D. A. Norman & S. W. Draper (Eds.), *User centered system design: New perspectives on human-computer interaction* (pp. 25-29). Hillsdale, NJ: Lawrence Erlbaum Associates.

Bannon, L. (1986b). Helping users help each other. In D. A. Norman & S. W. Draper (Eds.), *User centered system design: New perspectives on human-computer interaction* (pp. 399-410). Hillsdale, NJ: Lawrence Erlbaum Associates.

Bannon, L. & Bødker, S. (1989). Beyond the interface: Encountering artifacts in use. Aarhus, Denmark: Aarhus University, Computer Science Department. DAIMI PB-288. (To be published as a chapter in J. M. Carroll (Ed.), *Designing interaction: Psychological theory at the human-computer interface*).

Bannon, L. & Schmidt, K. (1989). CSCW: Four characters in search of a context. In *Proceedings of First European Conference on Computer Support for Cooperative Work* (pp. 358-372). Gatwick, UK.

Barnard, P. & Grudin, J. (1988). Command names. In M. Helander (Ed.), *Handbook of human-computer interaction* (pp. 237-255). Amsterdam: North-Holland.

Bewley, W., Roberts, T., Schroit, D., & Verplank, W. (1983). Human factors testing in the design of Xerox's 8010 "Star" office workstation. *Proceedings CHI'83 Conference on Human Factors in Computing Systems*. Boston, USA: ACM Press.

Bjerknes, G., Ehn, P., & Kyng, M. (1987). *Computers and democracy—a Scandinavian challenge*. Aldershot, UK: Avebury.

Bjørn-Andersen, N. (1988). Are "human factors" human? *The Computer Journal, 31* (5), 386-390.

Brown, J. S. (1986). From cognitive to social ergonomics and beyond. In D. A. Norman & S. W. Draper (Eds.), *User centered system design: New perspectives on human-computer interaction* (pp. 457-486). Hillsdale, NJ: Lawrence Erlbaum Associates.

Bødker, S. (1989). A human activity approach to user interfaces. *Human Computer Interaction, 4* (3), 171-195.

BYTE Interview (1983, February). An Interview with Wayne Rosing, Bruce Daniels, and Larry Tesler. *BYTE*, 90-114.

Card, S., Moran, T., & Newell, A. (1983). *The psychology of human-computer interaction.* Hillsdale, NJ: Lawrence Erlbaum Associates.

Carroll, J. M. (Ed.). (1987). *Interfacing thought: Cognitive aspects of human-computer interaction.* Cambridge, MA: Bradford/MIT Press.

Carroll, J. M. (1989). Evaluation, description and invention: Paradigms for human-computer interaction. In M. C. Yovits (Ed.), *Advances in Computers*, 29, 47-77. London: Academic Press.

Carroll, J. M. & Campbell, R. L. (1986). Softening up hard science: Reply to Newell and Card. *Human Computer Interaction*, 2, 227-249.

Carroll, J. M. & Mazur, S. A. (1986, November). LisaLearning. In *IEEE Computer, 19* (10), 35-49.

Carroll, J. M. & Rosson, M. B. (1987). The paradox of the active user. In J. M. Carroll (Ed.), *Interfacing Thought: Cognitive Aspects of Human-Computer Interaction.* Cambridge, MA: Bradford/ MIT Press.

Cypher, A. (1986). The structure of user's activities. In D. A. Norman & S. W. Draper (Eds.), *User centered system design New perspectives on human-computer interaction* (pp. 243-263). Hillsdale, NJ: Lawrence Erlbaum Associates.

Floyd, C. (1987). Outline of a paradigm change in software engineering. In G. Bjerknes, P. Ehn, & M. Kyng (Eds.), *Computers and democracy—a Scandinavian challenge* (pp. 191-212). Aldershot, UK: Avebury.

Giedion, S. (1969). *Mechanization takes command.* New York: W.W. Norton.

Gould, J. (1988). How to design usable systems. In M. Helander (Ed.), *Handbook of human-computer interaction* (pp. 757-790). Amsterdam: North-Holland.

Gray, W. D. & Atwood, M. E. (1988, October). Interfacing thought: Cognitive aspects of human-computer interaction [Review of *Interfacing thought: Cognitive aspects of human-computer interaction.*] *SIGCHI Bulletin*, 88-91.

Greif, I. (Ed.). (1988). *Computer-supported cooperative work: A book of readings*. San Mateo, CA: Morgan Kaufmann.

Helander, M. (Ed.). (1988). *Handbook of human-computer interaction*. Amsterdam: North-Holland.

Jones, J. C. (1988). Softecnica. In J. Thackara (Ed.), *Design after modernism* (pp. 216-226). London: Thames & Hudson.

Landauer, T. (1987). Relations between cognitive psychology and computer systems design. In J. M. Carroll (Ed.), *Interfacing thought: Cognitive aspects of human-computer interaction* (pp. 1-25). Cambridge, MA: Bradford/MIT Press.

Mead, G. H. (1934). *Mind, self and society*. Chicago: University of Chicago Press.

Newell, A. & Card, S. K. (1985). The prospects for psychological science in human-computer interaction. *Human Computer Interaction, 1*, 209-242.

Newell, A. & Card, S. K. (1986). Straightening out softening up: Response to Carroll and Campbell. *Human Computer Interaction, 2*, 251-267.

Norman, D. A. & Draper, S. W. (Eds.). (1986). *User centered system design: New perspectives on human-computer interaction*. Hillsdale, NJ: Lawrence Erlbaum Associates.

Reisner, P. (1987). Discussion: HCI—what it is and what research is needed. In Carroll, J. M. (Ed.), *Interfacing thought: Cognitive aspects of human-computer interaction*. Cambridge, MA: Bradford/MIT Press.

Shneiderman, B. (1987). *Designing the user interface: strategies for effective human-computer interaction*. Reading, MA: Addison-Wesley.

Suchman, L. & Wynn, E. (1984). Procedures and problems in the office. *Office: Technology and People, 2* (2), 133-154.

Thomas, E. L. (1965). Human factors in design. In M. Krampen (Ed.), *Design and planning*. New York: Hastings House.

Thomas, J. & Kellogg, W. (1989, January). Minimizing Ecological Gaps in User Interface Design. *IEEE Software*, 78-86.

Van Cott, H. & Kinkade, R. (Eds.). (1972). Human engineering guide to equipment design. Washington: American Institute for Research. (Revised edition).

Whiteside, J. & Wixon, D. (1987). Discussion: Improving human-computer interaction—a quest for cognitive science. In J. M. Carroll (Ed.), *Interfacing thought: Cognitive aspects of human-computer interaction* (pp. 337-352). Cambridge, MA: Bradford/-MIT Press.

Young, R. M., Green, T., & Simon, T. (1989). Programmable user models for predictive evaluation of user interface designs. In *Proceedings of CHI'89 Conference on Human Factors in Computing Systems* (pp. 15-19). New York: ACM Press.

3

Taking Practice Seriously

Eleanor Wynn

Sir, are you so grossly ignorant of human nature, as not to know that a man may be very sincere in good principles without having good practice?
> Samuel Johnson (Oxford Dictionary of Quotations, p. 272)

And it's not just the fact that you convince a customer that they owe the money, number one, like some of these people—which you know, is a big step right there. But then, to get it through their system, you have to understand what they have to do to get a check cut.
> Grace S., Billing and Collections Representative

Principles and Practice

Grace S. and Samuel Johnson, quoted above, are talking about the same thing, only Grace speaks from her experience at an administrative job in 1979, and Samuel comments from his observations in 1773. Grace finds that a principle—the principle that money is owed for goods delivered—is not a sufficient condition for the accomplishment of her task; she also must be able to manipulate practices: "You have to understand what they have to do to get a check cut." She goes on to illustrate that it is not simply a matter of knowing how that other system *ought* to work; what its principles are. Rather, in order to meet performance targets set for her (a certain amount collected by a certain date), she must use knowledge of general administrative procedure and her own organizational savvy.

> Once Accounts Payable gets it, then you get ahold [of the person there] and she'll say, "I have your invoice here, but it cannot be cut for two weeks." Your cut-off [target date] is one week. "Okay," you say, "may I speak to your supervisor, please," in a nice way, because you don't want to offend them to think that you, like [example of offensive version]: "MAY I SPEAK TO YOUR SUPERVISOR, PLEASE!"—make 'em feel like [implicit

communication of offensive version]: "because you are so incom-
petent," you know. What you say is, "Well, fine, thank you. I
understand that's your job, you know. Just—you have a job to
do like I do. I just want to speak to your supervisor so that
maybe we can work out something."

Grace's dialogue is taken from a long description illustrating how
her job requires skills that are not part of its formal description, and
which are recognized by her management only in their effects:
meeting the targets. The event she recounts involves an error in
billing that resulted in a delayed payment, and the complex sequence
of measures she took to accomplish a timely payment, using fine
points of organizational logic and skillful interpersonal maneuvers.

The quote is pertinent in bringing together her statement that real
performance in the job is by no means merely a matter of procedure.
She illustrates that social skills are critical to accomplishing the task.

Later in the chapter I will say more about the role of social knowledge and interactive skills in the workplace, but first I will clear up
some problems behind the chapter's deceptively simple title. It
refers to the way I, as an anthropologist, and others in the book
from their perspectives, regard the activities of people pursuing their
means of livelihood. What we do on the job is more than a rote performance or carrying out of simple "rules." It is a practice, a set of
skills, judgments, behaviors. It follows logically from that identification that one would take it seriously. It implies a stance not only
of responsibility toward the design undertaking, keeping in mind its
consequences and implications, but also of receptivity for the full
range of communications the practitioners have to offer about the
work in question. When we talk about "taking someone seriously"
it means we believe them and think they have something important
to say, not only in content but in framing information into a relevant
context.

Practice? Seriously?

But what is practice? And how do we go about taking it seriously?
These are tricky questions. In the quote from Johnson that opened
this chapter, practice is what a person does, as opposed to what a
person professes to believe. In Grace's context, it means all the
other things "you have to understand" in order to get things to work
the way they should, in principle, work on their own.

Practice is both "doing" and "understanding" that enables doing.
As used in the title, practice has a double—no, a single—well, a
double *and* a single meaning. It is what the system designer does at
the designing job, and what the prospective users do at the jobs

which, traditionally, the designer might transform with computer technology. Because it is what both kinds of people do at their jobs, it is a single thing. Because we make a distinction between the significance of the designing job and the job under design, practice, in practice, splits into two possible meanings: the practice of system design and the practice of everything else. Because the former can change the nature of the latter, it *is* in some way different. Because it is a job approached by people in the way people approach jobs in general, it is no different.

It is this "now you see it, now you don't" difference that makes the idea of taking practice seriously deceptively simple. Everyone who practices design thinks that he or she is taking practice seriously, so why is this an issue? I think most authors in the book are trying to say that too many system designers are taking *principles* seriously, and when it comes to someone else's practice, they don't really know how to get at it. In the system design situation, one of the inherent problems is that to take practice seriously as described here is to violate cherished premises of principle.

Principles are akin to the abstractions and structures that accompany doing science and mathematics. Practice is the more organic, messy, multi-dimensional reality—cognitive, social, goal-oriented, economic, mechanical, manual, emotional, verbal, these are some of the names given to the way we have separated those "dimensions"— of how things get done and make sense moment-to-moment, day-to-day at a job.

Practice in Philosophy & Sociology

Practice is more accurately stated as "practices." The term practices has a particular meaning in the context of a phenomenological view of human intelligence and action. The "background of practices" in which actions or utterances are situated—play a part, have a name, derive meaning—is the "ordinary behavior" of a group. This is what is assumed and expected, what is performed in the normal course of activity. It is the backdrop of things done routinely, against which certain things stand out as significant.

In his critique of artificial intelligence, and specifically of AI's attempt to codify human reasoning in computer-processable terms, Hubert Dreyfus (1979) writes about the background of practices as the source of sense-making that enables human beings to enact whatever "rules" might appear to operate in their behavior. This background can never be specified. It can be perceived in part, but not specified. This is the basis for Dreyfus' argument that machines can never think, or even be "intelligent."

My thesis, which owes a lot to Wittgenstein, is that whenever human behavior is analyzed in terms of rules, these rules must always contain a *ceteris paribus* condition, i.e., they apply "everything else being equal," and what "everything else" and "equal" means in any specific situation can never be fully spelled out without a regress. Moreover, this *ceteris paribus* condition points to a background of practices which are the condition of the possibility of all rulelike activity. In explaining our actions we must always sooner or later fall back on our everyday practices and simply say "this is what we do" or "that's what it is to be a human being." Thus in the last analysis all intelligibility and all intelligent behavior must be traced back to our sense of what we *are*, which is, according to this argument, necessarily, on pain of regress, something we can never explicitly *know*. (Dreyfus, pp. 56-57)

In sociology, closely related philosophical antecedents have informed an approach to the study of human behavior, mainly through examining language use and natural conversation. Here that background Dreyfus speaks of is, in a way, laid out for view in the assumptions embedded in everyday speech. Every conversation is premised on an assumed relationship and on presuppositions of shared knowledge between speakers. Speakers adapt to fill in where it is expected this sharing will be missing; or they leave out redundant reference when there is a large amount of experience in common. Since language has a certain "natural logic," and sentences have implications, we can observe a portion of the background of practices in what is left out and what is filled in. We can also note the categories and classifications people employ and how these affect action.

Within the field of sociology, this approach is called "ethnomethodology." The term refers to the way people see themselves in social action, their own structuring (methodology) of their ways (ethno-). Harold Garfinkel (1974), who first used this approach, describes it as "the organizational study of a member's knowledge of his ordinary affairs, of his own organized enterprises, where that knowledge is treated by us as part of the same setting that it also makes orderable" (p. 18).

This is related to theories of social action and social constructionism in saying that people are constantly engaged in the process of creating and reaffirming the social order. Content is a function of this process. Many social researchers and system designers are interested in "content," as if it could mean something on its own.

Another term for the activity of making sense of one's world is *practical reasoning*.

The focus on practical reasoning emphasizes that the talk accomplishes scenes and their contained activities; it emphasizes that members are—as a condition of their competence—rendering scenes intelligible, reasonable, and accountable, that their world is a constant *doing* and *achieving*. "Practical" actors make and find a reasonable world; their doing so is topically available for the social scientist. (Turner, 1974, p. 10)

Are You Talking to Me?

But what does this have to do with system design? System designers are not studying the nature of human intelligence, nor how society is put together. They are simply trying to understand how a particular job is done so that a computer program to use on that job will be an improvement, useful, appropriate, and not in the way.

There is a problem. System designers are forced to be students of human intelligence and society when they set out to understand a set of human activities organized into a process of production. Their normal training has made them good at manipulating symbols or materials that have predictable properties under specific conditions, that can be formulaically arranged to produce a certain outcome. The preceding argument is that the workplace, being a socially created and maintained place, doesn't consist exclusively of such elements. It is based on this background of practices, in fact, practices within practices. These encompass everything from management style and industry conditions to the logistics and rationales of departmental functions, reward systems, territorial and political concerns among departments, the procedures and skills of the work, and the basic social competence of the participants.

Much of this does not fit into the context of the common system design methods, even if it could be articulated. Imagine asking someone how he swims. Even if he could describe the movements adequately, he probably would not mention the pre-condition: you have to be in water, alive.

System Design Is a Social Encounter

Furthermore, most workplace situations we encounter in doing either research or design have a place in our own generalized background. We have preexisting categories for the people we encounter and the activities we observe. There is a taken-for-granted aspect to much of it, against which we place our particular agenda.

There are elements of this general background that may block out important perceptions and communications, on both sides. On their end, too, the participants may have us categorized. They may have opinions about what we are really up to: snooping, wasting time, evaluating, performing wizardry, making a brief descent from the ivory tower to peer at real work. It is ridiculous to approach this as if it weren't first and foremost a social encounter, which contextualizes and determines the precise information that will be produced. But the participants will usually be able to see us more as we are, if we can see them more completely as they are. More equally. Less stereotypically. That is an important part of what it means to take practice seriously.

This is a delicate thing to say, because it implies that we have been violating basic social principles. Almost everyone at least pays lip service to egalitarian social beliefs; in fact, most probably believe that they have "good practice" in that regard. Yet it is part of the unarticulated background to divide people up into social categories. This is natural and in itself not a problem, because there *are* differences in the speech, behavior, tastes, and knowledge of groupings of people; these often correspond with other socioeconomic distinctions. The problems come with exaggerations that accompany social stereotypes.

Social Distance

I myself observed such stereotyping frequently during the mid-1970s when the application and usability of "electronic office systems" was an issue in computer research and development organizations. Managers of business functions and research scientists each in their own way made suggestions about user interfaces, training, and applications design that reflected the assumption that endusers—clerical in this case—were neither intelligent nor adaptive.

Each group, line management and technical designers, had its own distancing mechanism that prevented them from seeing the working person very accurately: organizational role in the former, and technical education in the latter case. More to the point, they seemed unable to pay close attention to the speech or activities of the potential users, or to suppose that their discourse could be taken at face value or contain complex information. It was as if the clerical people were being viewed through the wrong end of the telescope; when they were so close at hand, no telescope was needed.

Social boundaries are a normal feature of human societies, for better or worse. They serve a healthy function in providing meaningful identities and closely woven backgrounds of practices for their members, so that there is predictability, form, and order within groups as well as across. They pose problems in terms of our contemporary values when they rigidly support and enforce systems of social stratification. The effect of boundaries is a greater or lesser social distance. Certain activities are done within the group, others between groups.

Social distance can be maintained without physical distance, in the form of stereotypes and assumptions. Real differences such as education, income, and social dialect may be taken to imply differences that are not real in intelligence, ability, responsibility, credibility. In the terminology of Marx and Lukàcs, these added-on evaluations are part of a construct called "false consciousness." As a deficient form of class consciousness—the ability of a socioeconomic class to see itself as having a common political interest—false consciousness describes a group's ability to see itself as a class but without perceiving its relationship to the means of production. Instead it justifies a social order of inequality by attributing unfavorable attributes to the less favored classes.

Again, what does this have to do with system design? A subtle but pervasive false consciousness inhibits many encounters between people of different backgrounds. It affects the assumptions we hold about others, and hence affects our interaction and observation. It therefore affects what we learn about both them and the situations they are in. A technical education, and how a person came to have that education, forms a specific background.

Scientific Detachment

In fact, any form of scientific education builds another boundary between researchers or designers and the participants in the workplace. C. Wright Mills (1959), in his classic treatise on the practice of sociology, denounces the "false consciousness of scientific detachment." He was referring to the social sciences and the preference for statistical rigor over meaningful issues, the dominance of the "intellectual technician" as opposed to the "intellectual craftsman." Mills was critical of the insistence of many sociologists on being seen as "hard scientists," modeling their work after the physical sciences. Scientific detachment in this context is a pretense, since:

> Values are involved in the selection of the problems we study; values are also involved in certain of the key conceptions we use in our formulation of these problems, and values affect the course of their solution.

> The confusion in the social sciences is moral as well as "scientific," political as well as intellectual. Attempts to ignore this fact are among the reasons for the continuing confusion. In order to judge the problems and methods of various schools of social science, we must make up our minds about a great many political values as well as intellectual issues, for we cannot very well state any problem until we know *whose* problem it is. (pp. 76-78)

Scientific detachment is also a problem in the physical sciences, although in a slightly different way. There, values are also involved in the selection of problems and perception of subjects, but it can be argued that values are not inherent in the subject matter itself, as they are in the social sciences. Still, the fact is that most system designers draw on their mathematical/scientific background for ways of viewing the world, and scientific detachment is part of that.

I said earlier that doing system design is like doing any job. It also relies on an unarticulated "background of practices." Thomas Kuhn (1970) has developed this theme most thoroughly in *The Structure of Scientific Revolutions*. In discussing how "paradigms could determine normal science without the intervention of discoverable rules," he gives the following account of the scientist's backgrounded model.

> First ... is the severe difficulty of discovering the rules that have guided particular normal-scientific traditions. The difficulty is very nearly the same as the one the philosopher encounters when he tries to say what all games have in common [referring to Wittgenstein]. The second, to which the first is really a corollary, is rooted in the nature of scientific education. Scientists ... never learn concepts, laws and theories in the abstract and by themselves. Instead, these intellectual tools are from the start encountered in a historically and pedagogically prior unit that displays them with and through their applications. (p. 46)

Perhaps even more to the point for this book, which proposes a different approach to system design appropriate to the context of human practice, is Kuhn's observation that

To the extent that normal research work can be conducted by using the paradigm as a model, rules and assumptions need not be made explicit .... But that is not to say that the search for assumptions (even for non-existing ones) cannot be a way to weaken the grip of a tradition upon the mind and to suggest the basis for a new one. (p. 88)

Rather than being validated exclusively in method, scientific work is more realistically

honoured by the relevant colleague group as "normal" or "competent"... [it] is through-and-through a derivative of common sense. Scientists rely upon a "syntax" of practices and methods which are accredited as "correct," "sufficient to the task at hand," "properly conducted by prevailing standards" in just those ways in which any concerted activities are warranted by a collectivity. (Turner, 1974, p. 8)

The point of the foregoing is not to discredit traditional and "scientific" approaches to research or design, but to say that everyone brings to a task a set of presuppositions, not necessarily grounded in anything other than practice. Scientific and mathematical practitioners are no exception. Moreover, as the problems in the field change, that set of presuppositions may no longer be adequate. Instead, it blocks the way to a more inclusive paradigm. Now it seems that invisible assumptions in system development, both social and methodological in nature, stand in the way of creative innovation during a time of pervasive technological change.

The accepted forms of the past offer a sense of security both about the outcome of a project as a professional artifact, and about its acceptance by the "relevant colleague group." But what about the outcome for the users? The recent history of computerization provides many instances of programs that don't work as expected, even though they faithfully execute the description that was obtained.

In the January 1990 *Communications of the ACM*, there are no fewer than three editorials that address this subject. Two of them highlight the irony, not to say tragedy, that the consequence of poor design is not just that systems fail to perform properly, but that "human error" is invariably cited as the cause! The human in question is the hapless person at the controls during a crisis, not the team of humans who designed the system, nor the human relations that set the context for the design interviews.

Many advances have been made in our understanding of the hardware and software of information processing systems, but one major gap remains: the inclusion of the human operator into the system analysis ... The designer must consider the properties of all the system components—including the humans—as well as their interactions. (Norman, 1990, p. 4)

While this appeal is absolutely correct in its import, the roots of the problem are carried over into the proposed solution—in the way the exhortation is phrased. First of all, reference to "the human operator" implies that it is always one-on-one: one person-one computer. We have talked about people in social groups, applying group processes and group values to their tasks. The use of the term "operator," even more than the term "user," subordinates the person, by implication, to a machine-tending rather than a formative role. The call to "consider the properties of all the system components, including the humans," unwittingly reduces people to components with properties. In an earlier section I posed the idea that the "background of practices" is hard for technical people to consider because they are more accustomed to looking for "properties."

If one of the more sensitive and practical members of the computer establishment can, with the best of intentions, present so limited a view of what might be involved in including humans, then clearly there is much soul-searching to be done. Otherwise, the properties will be studied, the systems will still have problems, and the humans will be blamed all the more because they were included. It isn't just *the* operator, and it isn't just in relation to the system, but to a whole world. This world can't be understood by sitting the "operator" at a console in a laboratory, as Liam Bannon explained in Chapter 2.

Can We Talk?

The important thing about the background of practices for our concerns in this book is that they are the product (and process) of a group. In different ways, the various authors in the book have either observed, or participated in, groups in such a way as to be able to be aware of that background. In anthropology, the term "participant observation" is used to describe the basic approach of the field. It refers to extended participation, as an outsider, in a group being studied; being involved so as to fit in, while noticing and documenting all along. The methods used throughout the book tend to differ from that basic approach but to incorporate key features of it.

This method of participant observation makes anthropology distinct as a social science and makes its approach uniquely useful in contemporary situations of technological change. First, it is oriented toward understanding whole life systems, starting with the basics of livelihood and ecology. Second, its practice necessitates being close, to a functional extent, with the people being studied. In the past, anthropologists spent years living among, and hence depending upon their social relationships with, another people for both their physical existence and the information they came to get.

This has always made "scientific detachment" problematic in anthropology, but it has also produced a social science that is accountable to its subjects as much as to its own informing theories. It gives the people of the study a chance to speak for themselves, to draw their own picture. There is no prior decision to include some kinds of information and exclude others.

The analysis unfolds or emerges from connected sequences of naturally-occurring behavior or talk. The work can be daunting, in that there is an overwhelming richness of information. Much of it is in the form of detailed sequences of micro-scaled enactments of what can only later be called "structures" of speech or behavior. The data may seem formless at first. It has to be combed over repeatedly. The analysis emerges from the data, which tend to be voluminous and available to several different levels of examination, being as they are excerpts from ongoing real events in the life of a group.

In the case of observation within our own main culture, there is the additional difficulty that much of the behavior is "transparent" to us. It looks normal and routine. It is part of our own background. The researcher or designer must in a way reverse the field, pull the background into the foreground, and begin to see how portions of behavior function as a part of the process. In the case of workplace studies, this means seeing how those unarticulated or "glossed" practices support the work.

One Person's Experience

My own work in this area began in 1974, when I decided to write a doctoral dissertation on office conversation and how it reflects the hierarchy of authority in an organization. In 1976, I held a summer internship at Xerox Palo Alto Research Center. During the course of that summer, my interest shifted from the hierarchical aspect of office talk to the informational aspect. How does the social environment at work support information? In a study of a secretary's conversation, I found the social and information elements to be blended together. In other words, there appeared a reason to think that the social environment of a workplace supports more than camaraderie and good feelings.

I went on to do a larger study of clerical workers in a highly structured situation; that of a telephone sales order entry organization. This environment consisted of employees who received calls from customers wishing to order copier supplies, and a companion group that adjusted customer accounts in the case of errors in billing or shipping, returned goods, etc. Here the pace of work was much more intense than at the desk of the secretary from the preliminary study. Nevertheless, there was continuous talk, both between the employees and the customers and among the employees (Wynn, 1979).

Work conversations dominated the 25 hours of tape recordings that I used for my study. But the conversation here was no less a product of the social context than was the secretary's. Again, social bonding cues and task information were wound together, as part of the same social competence. This data made a strong case, I felt, for the critical importance of the social environment as an information resource for employees. The observation that social networking supports knowledge activities had been made in the past for research professionals, but it had never been extended to clerical people, whose work was generally seen as routine, form-based, and procedure-driven.

Contrary to this supposition, I found the following information activities recurring throughout the data:

- defining procedural terms
- categorizing products, situations, problems
- establishing relevance of facts to a situation
- providing historical context for interpretation
- situational pattern-matching, for clues to a problem
- conjecture, explanation, clarification, verification
- instructing other employees
- social bonding with customers during problem-search
- generally maintaining cooperative ambience

These activities had to be fished out of the ongoing flow of talk. But once I began to "see" these functions, I found them permeating the talk.

One reason why taking practice seriously contrasts with existing design approaches is that "practice" incorporates the whole of the working person's social presence. The person who works with information deals with an "object" that is more difficult to define and capture than information flow charts would have us imagine. These show "information" in little blocks or triangles moving along arrows to encounter specific transformations and directions along the diagram. In reality, it seems, all along the arrows, as well as at the nodes, that there are people helping this block to be what it needs to be—to name it, put it under the heading where it will be seen as a recognizable variant, deciding whether to leave it in or take it out, whom to convey it to. The information doesn't come in from the outside already formatted for processing in many of the cases. Even when people perform seemingly routine functions with information, ambiguity and uncertainty must be handled before the process can continue.

I found the work group continually applying this added value, as a result of being a social group in verbal contact. This is performed as a background activity, not as something explicitly noticed. A more traditional approach would treat both the information and the functions as if they were permanently defined, always recognizable entities. But this definition and recognition are in a sense *accomplished* by workers as a part of their general, taken-for-granted social competence. As a feature of social competence, it is carried along the same "channel" as other social activity.

Paul: Hello, Jan?

Jan: Mhmmm?

Paul: I've got a gal who got some bad—boxes of labels.

Jan: I heard ya tryin'a talk 'er out of 'em.

Paul: Yeah, they're curlin' up and such, on the sides.

Jan: Arright—just take the—d'y'have the order number that she ordered 'em on?

Paul: I've got—no. She, I'm getting 'er—she's got the old invoice. It was back in March, 'n this doesn't go back quite—it goes to March, but not when she ordered 'em.

Jan: Okay, arright.

Paul: So, I'm gonna get that invoice number that we billed 'er on. And then, uh—

Jan: Get the supply order number—yeah, get the *supply* order number off that invoice 'n then I c'n send 'er somp'n.

Things to note in the above conversation are that:

1. Paul doesn't have exactly the documentation he needs, and the record he is looking at cuts off at about the point where the number he needs would be located.

2. He has a strategy, which Jan either augments or confirms, to find an alternate document and take a number off of it.

3. Everything about the conversational style works to personalize: "I've got a gal who got some bad boxes;" Jan's remark about "tryin'a talk 'er out of 'em;" "they're curlin' up and such;"

always referring to the customer as "her;" and of course the pervasive use of informal articulation.

In this, as in all of the conversations, there is a mix of cooperative problem-solving and an informal, comradely way of talking that includes humor, irony, and folksy figures of speech. The implications of such linguistic features have been discussed thoroughly in the works of sociolinguists and linguistic anthropologists, notably John Gumperz (1971). Georg Simmel also noted that "idle talk" serves to maintain social channels of communication, even when there are no "important" things to communicate (Wolff, 1950). Similarly, linguistic features of camaraderie maintain a general atmosphere of sociability which facilitates routine and constant encounters over small problems.

In the customer service job, which involves resolving discrepancies over the telephone, this sociability is even more pronounced. In these situations, there is always the question of who will essentially have to pay for an error. The resolution depends in part on documentation, in part on policy. Resolution of the question relies heavily upon the historical precedent recollection, procedural knowledge, and general problem-solving skills of the customer service representative. In my tapes, the lead representative was also sensitive to her role in smoothing over any doubts the customer might have about getting a refund, while she worked to figure out the particulars of the case.

Jan: (on the phone with a customer) Yes, Ann. ... O-okay, mhm. Sure, can do. Can I have, d'you have the green packing slip, uh, in front of you there?... Okay, Ann, that, uh, "E" number? ... UHUH! Oka-ay. An' the name of the company, Ann?... Okay, have a li'l, little misspelling, huh? We'll see if we can get that corrected. An' gimme that customer number that's, uh, at the very top, Ann. ... Do these boxes look a little different to you from the ones you normally receive? In other words, possibly they don't have the regular typewritten numbers—maybe somebody's written the number in?... I have a suspicion they're repacks, oka-ay?... Possibly ... the rest is OK, Ann? OK. ... Well, what I'm trying to determine, is—I'm 'onna send ya, whatever y'need, y'know, an' no problem. I just wanted to make sure that none of the other ones were damaged, or had anything wrong with 'em .... Yeah ... yeah, that's understandable. How many total were there, Ann?

In this conversation, Jan accomplishes a good deal with emphasis, with her frequent calls for confirmation from the customer—"oka-ay?"—and with her constant use of the customer's first name. Also typical is the way she guides the customer to look at documents for her, so that she can get the clues she needs about the problem. Her accommodating stance is also shown in affirmations of service: "Have a li'l, little misspelling, huh? We'll see if we can get that corrected;" "I'm 'onna send ya, whatever y'need, y'know, an' no problem."

It is doubtful that Jan received any instruction on this aspect of her task beyond, possibly, some general indication to "be pleasant."

The particular ways that both the foregoing conversations implement sociability might be noted by system designers and others who interview people at work. While the particular speech styles will differ, a resort to informal speech seems to be used by workers in the U.S. with each other and with customers as a way of indicating social closeness or minimizing social distance.

Another important skill is the social competence involved in perceiving and defining situations. A situation is purely a socially created event, in the sense that without values, expectations, and consequences, all socially determined, it would be impossible to *have* something called a situation. A situation stands out against the background of practices as a story, a problem, a conflict. Incorporated into a situation is the possibility of overlapping worlds of values, such as the values in the logic of internal procedures, and the expectations of other departments, or customers.

Sue: This is Sylvia in—in Alaska [on the phone].

Joan: Alaska?

Sue: And still a problem at the U of Alaska. She's asking if we—could not get the proof of delivery, are we going to credit this customer out? Isn't our contract account involved? And we show it was shipped twice.

Joan: Shipped twice.

Sue: On the same PO number.

Joan: Uhuh. A duplicate shipment.

Sue: Yes. That it definitely was.

Joan: OK, is Linda M going to be able to give proofs of deliveries?

Sue: Well, she's asking me if, uh, well, I guess she [Sylvia] is going to call Linda about it. Cause she can't find a copy of it.

Joan: Linda can't?

Sue: No.

> Joan: I think you better call Linda. Rather than Sylvia calling
> her. 'Cause then you c'n tell 'er exactly what you need.
> They're going to have to end up crediting if she can't get a
> copy of the proofs of deliveries. So, tell 'er you'll—you'll
> contact Linda and get back to her [Sylvia]. We don't know
> if we're going to credit an amount till we 'ave proofs of
> deliveries.

Again, this is just one of a constant series of instances involving
situations that come up regarding discrepancies between what ought
to happen and how it works out. The internal documentation shows
there was a duplicate shipment, but the documents proving the
shipment was delivered are missing. Sue, in her presentation, goes
through the pertinent details as she sees them, then raises the key
questions: "Are we going to credit this customer out? Isn't our
contract account involved?" This basically states the two conflicting
elements of the situation. Probably the credit is in principle un-
warranted, but in practice they will have to do it. The contract
account indicates the customer's relationship—they have agreed to
purchase so much in supplies in order to get a certain price. This
relationship is also something to consider. But the credit without
full documentation will probably detract from the department's
performance measures.

In the early part of the conversation, Joan echoes the pertinent
facts and definitions. Her solution is focused on getting the documentation that they need to avoid crediting. Here she brings in an
interpersonal strategy, not unlike the approach Grace S., in our
introduction, uses when the chips are down. Get on the phone
yourself, "then you c'an tell 'er exactly what you need." You might
also get her to try harder to find it.

Sociologist Don Zimmerman (1974), looking at social competence
in pulling together adequate documentary cases, wrote the following
about the bureaucracy he examined:

> The persistent theme emerges...as the continual interplay between
> the routine and the problematic, the taken-for-granted use of
> documents and the occasioned accounts which make their use
> observable as rational procedure.

> The taken-for-granted use of documents...is dependent on the
> document user's sense of an ordered world of organizations, and
> an ordered world of the society-at-large. Indeed, such routine
> recognition, and the inference proceeding from it, is the mark of
> the competent worker.

> When a document is rendered problematic in a given case, the
> document-producing activities of which it is a part are made ac-

countable as orders of necessary motives, necessary actions, and necessary procedures which may be used to analyze the features of the case and reach a determinate and warrantable decision. It is the artful accomplishment of personnel that they are able to provide such accounts which sustain the organizationally required use of documents over the manifold contingencies of everyday investigations. (p. 143)

People, Not Procedures, Make Sense

One feature of organizations that rarely comes across in structured interviewing is the chaotic element that enters in when management changes, organizations are restructured or realigned, functions are cut or added, and of course, automated systems are introduced. The facade, the public image that any organization projects to outsiders is one of order, conformity, a smooth integration of procedures and functions from one location or department to another. True, the whole system tends, one way or another, to work in most cases, but not necessarily because of the idealized setup that is offered to the outsider as a description.

Not only must people bridge this gap between idea and reality, they also must at times suffer the consequences. When organizations don't "make sense," the people in them are aware of this, because they themselves work to create a framework of sense-making. Systems that violate this, whether new procedures or new software, create a serious gap in people's feeling of living in "an ordered world of organizations." Grace S, in our interview for one study I did, was brooding over a perplexing incident.

EW: What's happened?

GS: Ohh, everything. Well, we just got another memo which said they were going to change, uh, one of our policies.... They're not really gonna change it but they're just emphasizing to use it correctly.... Yet they really hadn't taken in the practicality of the total memo and what actually they're saying. So it creates a lot of frustration. Because we're trying to practically implement something that's written on paper, that when we call Region [management center] to get it, uh, explained further, ... it was almost a humorous response that we got back....They want us to adhere to this policy that we just received today... [that had the effect]—they really made it impossible to get rid of ... [a debit of] twenty-eight thousand dollars off our account. [which was not in fact a real debit]. How are we supposed to clean it off if we cannot use the credit that they, you

> know it was specifically sent for this—balance. And
> they're [Region] saying *impasse* it. "Let the branch eat
> twenty-eight thousand dollars."

Here Grace's sense of "an ordered world of organizations" has been completely violated. Because in the terms that count the most, the proverbial "bottom line," the branch (her location) has been asked to act as if they took a loss of $28,000, when in fact they didn't. And the reason they are asked to do this is to maintain a particular accounting format that is suddenly being enforced. Procedures, not uncommonly, violate the most basic common sense. Their enforcement is not consistent.

Again, the world of human activities, with all its complexity, cannot necessarily be expected to run on the schedule idealized in procedure. Yet the formal requirements must somehow be met.

> CB: And I plan my time around my calendar, and then adjust it
>   according to the fire drills that come up.
>
> EW: Do they come up a lot?
>
> CB: Sure do...all year long has been one huge fire drill....We
>   had a month-long fire drill on budgeting, because we
>   didn't have time to do it well the first time. Managers were
>   on vacation when the deadline was given to us by
>   [headquarters], okay? And I spent about a week doing a
>   dummy budget for the full year budget. And this was like,
>   each territory—there's sixty territories—each one has nine
>   categories to be budgeted. Well, I did a dummy budget
>   and we sent it in with a letter saying please give us an
>   extension, meanwhile here's our, you know, to show our
>   heart's in the right place, this dummy budget. Which they
>   then proceed to punch in. Erroneously. And then they
>   came back rejecting our request for more time.

I include these last two sections specifically to caution that the elicitation of a description of procedure—of how a manager idealizes the organization's functions—has a good chance of missing key elements of the real way things are done.

Now What?

Now, how can this understanding be incorporated in the designs of the future? Early on I grappled with this question of "given what you have found out, what do we as designers do?" I finally realized this is the wrong question. Anyway, I wish to reframe it. Really the finding is an ontological one, about how people *are*. Perhaps the shift in design practice also is more of a way to *be* than a thing to *do*.

The system designers who contributed to this book talk about some interesting things to *do*. What is interesting about these things to do is that they are *processes* that imply a way to *be* with respect to the users. Just as users are involved in their worlds of work in a whole way, designers also need to be involved with users in a whole way, as people. That is the only way to learn from them what they really know. Involvement is the opposite of "scientific detachment." That is why I said in the beginning that to take practice seriously may violate premises of principle.

Linguistic and social science research like the work reported in this volume ought to enable designers to *be* more sensitive, to *be* on the alert for cues to the nature of the organization as a whole, rather than just rely on its formal description. Part of competence is knowing what to say in which conversation. If the designer can succeed in making the conversation *be* a more open and general one, then it is possible to get a description of the way the work *is* rather than the way it *ought* to be. Similarly in observing, the designer can refocus his or her view to take in those other levels.

Another way to be is more trusting, both of the competence of the users to provide a meaningful picture, and of one's own competence to make sense out of a less structured, more diverse mass of data; to tolerate a temporary confusion in perusing this mass until the picture emerges.

Returning to our opening quotes, so different in time and context, so similar in point of view, let's be serious about having good practice, along with being very sincere in good principles. And, like Grace, let's make sure we "understand what they have to do to get a check cut." At this point, the technology, the programming languages, the sophisticated system concepts, are more than adequate to the purpose. It's the purpose itself we need to comprehend. This comes through the logic of conversation, the logic of situations, the logic of involvement—not mathematical logic. It's hard to let go of mastery. But there is an obsessive quality in the repetitious exercise of something one does to perfection at the expense of communicating with the person nearby, even in an imperfect way.

One result of being more involved with users is that the designer will inevitably begin to feel, to be, more responsible to them and to how the system will accommodate them as people, not just as an abstraction of a work process.

Like the users, who are practical reasoners, the designer can take a practical stance toward the setting rather than one dictated by a methodology.

References

Dreyfus, H. L. (1979). *What computers can't do* (rev. ed.). New York: Harper & Row.

Garfinkel, H. (1974). In R. Turner (Ed.), *Ethnomethodology* (p. 18). Baltimore: Penguin.

Gumperz, J. J. (1971). *Language in social groups.* Stanford, CA: Stanford University Press.

Kuhn, T. (1970). *The structure of scientific revolutions* (2nd. ed.). Chicago: University of Chicago Press.

Meszaros, I. (Ed.). (1971). *Aspects of history and class consciousness.* London: Routledge & Kegan Paul.

Mills, C. W. (1959). *The sociological imagination.* London: Oxford University Press.

Norman, D. A. (1990, January). Viewpoint. *Communications of the ACM, 33* (1), 4-7.

Oxford Dictionary of Quotations (3rd ed.) (1980). New York: Oxford University Press.

Turner, R. (Ed.). (1974). *Ethnomethodology.* Baltimore: Penguin.

Wolff, K. H. (1950). *The sociology of Georg Simmel.* New York: Free Press.

Wynn, E. H. (1979). *Office conversation as an information medium.* Doctoral Dissertation, University of California, Berkeley.

Zimmerman, D. H. (1974). Fact as a practical accomplishment. In R. Turner (Ed.), *Ethnomethodology* (p. 143). Baltimore: Penguin.

4

Understanding Practice: Video as a Medium for Reflection and Design

Lucy A. Suchman and Randall H. Trigg

Work as Situated Activity

In exploring the design of new technologies at work, we begin with the view that work is a form of situated activity. By this we mean that work activities in every case take place at particular times, in particular places, and in relation to specific social and technological circumstances. From this perspective, the organization of work is a complex, ongoing interaction of people with each other and with the technologies that are available to them.

Because of the intimate relation between work and technology, the development of the artifacts with which people work and the development of their work practices go hand in hand. Available technologies afford certain resources and constraints on how the work gets done, and peoples' ways of working give the technologies their shape and significance. Our research goal has been to understand that work/technology relation both more generally, developing theoretical constructs that can deepen our understanding across work settings, and more specifically, in terms of the detail of just how the relationship develops in and through the work's course in a particular setting.

Routine Trouble in an Airline Operations Room

We are currently engaged in a project to study the relation between facilities design and everyday work practice. Our goal in the Workplace Project is to understand design effectiveness from both designers' and users' points of view, through an intensive field study of a complex work setting. After extended deliberations, we decided to locate our three-year study at a local airport. The airport is particularly interesting from our perspective in that during the course of our study a new terminal will open. This will provide us the opportunity to look at the process of design, use and redesign in a kind of accelerated time frame, as people move their operations from facilities whose design has been developed, modified, and adapted to over time, into new facilities based on projections of use which we assume will be more and less accurate. Our first task, therefore, has been to look as closely as we can at the organization of work operations in the existing terminal.

Figure 1a. The "Ops Room."

During one evening of fieldwork at an airline operations room, we observed the management of some "routine trouble" through the deployment of a variety of artifacts and communications technologies. Before exploring the details of the incident, some background on airport operations and the particular work of this room is in order.

Figure 1b. The "Ops Room."

The operations room is a communications center that coordinates the ground operations of a single airline (see Figures 1a and 1b). In this "Ops Room," staff consists of five airline employees. Each has specific responsibilities relevant to getting planes into and out of gates and to transferring baggage and passengers between planes. Their efforts are especially concerted during what are called "complexes." These are periods lasting approximately an hour, when all of the gates belonging to the airline fill with incoming planes, transfers are made between the gates, and then all of the planes depart. There are eight complexes in a normal workday. On the evening in question, the Ops Room was managing a multiple airplane "swap" which, among other maneuvers, involves one airplane arriving at gate 18 during complex 7 and departing from gate 14 during complex 8, an hour or two later.

The information necessary to coordinate the work of a complex is present both online (in an airline-wide computer system) and on paper documents called "complex sheets" (see Figure 2). These sheets contain matrices mapping incoming to outgoing planes, and include cells for each transfer of people and baggage required during the time the complex is on the ground. The dynamic nature of the complex is captured on the sheet by ordering the rows and columns of the matrix chronologically. Thus an Ops Room member checking off completed transfers should generally be moving diagonally downward and to the right across the cells of the matrix. Delayed flights display themselves as groups of cells left behind in this process.

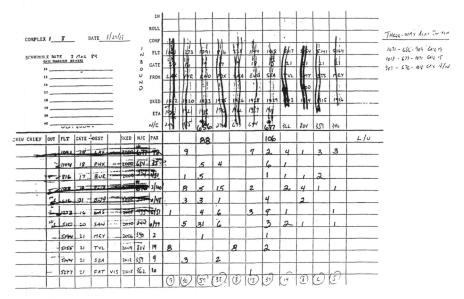

Figure 2. The complex sheet.

The dynamics of the complex in question, however, present problems for these established procedures. The major problem is that the aircraft swap takes place across two successive complexes, and thus stretches the complex sheet design in various ways.

In the interaction transcribed below, we focus on the activities of two Ops Room employees: BP (Baggage Planner) and PP (Passenger Planner). For our purposes, it's enough to say that BP coordinates and communicates with crew chiefs on the ramp, while PP plays a similar role with respect to ticket and gate agents inside the terminal. In the first sequence of the interaction, BP approaches PP with a question about tonight's plane swap (Figure 1a). PP provides an explanation and volunteers that perhaps he should have notated the complex sheet differently (Figure 1b):[1]

5:49:34 pm

((BP sitting at her workstation looks over her shoulder.))

BP: Dave? (2.0) ((Gets up with clipboard and complex sheet and walks over to stand beside PP)) I'm sorry to keep bugging you I just want to make sure I figure this out.

PP: Okay. <inaudible>

BP: Nine-oh-nine inbound:

PP: Takes out ten-eighteen=

BP: =Ten-eighteen, but you've got gate eighteen?

PP: Uh, okay, here's the problem.

BP: Isn't that supposed to be

PP: He comes into eighteen, but then eh, we're going to move him off of eighteen to fourteen so this ((reference to complex sheet)) is right.

BP: Okay, that's right. Okay, that's r: 's the one they're

PP: This here plane is going to be moved after that seven o'clock departures, uh, complex seven goes out.

BP: Okay, right.

PP: So, I should have put eighteen slash fourteen on there is what I should have done.

BP: Right, okay, yes I saw fourteen and then eighteen. Okay, I got it. ((Goes back to her desk and sits down.))

The complication here is that flight 909 comes in to gate 18 during complex 7, the aircraft is moved to gate 14, and it then departs, during complex 8, as flight 1018. The complex sheet, however, shows only that Flight 909 comes in to gate 18, leaving it unclear as to how flight 1018 could then leave from gate 14. The difficulty arises from the fact that the complex sheet a) is not designed to show the movement of aircraft and b) is designed to cover only one complex at a time. The solution devised by PP, refined in this exchange with BP, is to annotate the sheet to show the plane "swap" across the two complexes.

---

[1] Our transcription conventions are derived from those of Jefferson (1984). (A condensed glossary of Jefferson's transcription symbols can be found in Heritage (1984), pp. 312-314). In the transcripts, equal signs (=) indicate "latching," i.e. the beginning of one utterance following directly on the end of the prior with no gap. Numbers in parentheses indicate elapsed time in seconds. Thus (2.0) indicates a pause of two seconds. Colons indicate prolongation of the immediately preceding sound.

Almost immediately following this interchange, PP decides to check the online information to make sure that it correctly (and readably) reflects the state of the world:

> (3.0)
>
> PP: Think I put that in the oh-ess-oh-star ((a program on the airline computer system)), I better double check that. ((Sits down at computer and begins typing.))

Having clarified the online record, PP makes two announcements over the radio, each designed to help the gate agents make sense of the situation. The first encourages them to check their computers for the latest information. The second proposes the same modification to their copies of the complex sheet that was arrived at earlier with BP.

> 6:00:23 pm
>
> PP: ((Leaning over to speak into radio)) Once again, just a reminder to everybody, make sure you check complex seven and eight in the oh-ess-oh-star. There's a lot of information in there. It might be a little bit, uh, hard to understand on that ten-eighteen outbound, that airplane is gonna come in on gate eighteen and be moved to fourteen. So if there's any question on the gate for ten-eighteen outbound, that departure will be gate fourteen.
>
> (2.0)
>
> Radio: ((Female voice)) Thank you David I was about to call you about that.
>
> (2.0)
>
> PP: ((into the radio)) Yeah, Crystal and for everybody else nine-oh-nine inbound is the aircraft that makes up that ten-eighteen and when he comes in, he parks at eighteen but, due to the problem tonight with all these changes we'll have to move that aircraft over to fourteen, after the seven o'clock departure.
>
> (10)
>
> PP: ((again into the radio)) On that complex seven cut sheet uh, if it would make it any easier, where it says that three-way aircraft switch up there in the corner (.5) the third flight number listed, nine-oh-nine in, ten-eighteen out, says gate eighteen, put a slash behind that put fourteen behind it.

Using the Complex Sheet

Now suppose that we are interested in exploring new technologies for this workplace. And suppose further that the complex sheet is being considered for replacement or augmentation. Careful reflection on the activities of using complex sheets might first make clear the multiple purposes for which the complex sheet is used, so we can then ask how well those purposes are served by its current design. We have found that the complex sheet acts as

1. A *reproducible representation* of the intended motion of people and baggage over the course of some complex. Note that we're not saying it's a set of instructions; its users generally know exactly what to do, they just need access to the projected particulars of a given complex;

2. A *template* which is filled in, in various ways, depending on features of the particular complex and on needs of the user. In fact this templating happens at several levels. First, the matrix format (layout, number of rows and columns, etc.) is a template onto which the projected incoming and outgoing flight numbers and arrival/departure times are notated. This happens every morning, so that when they arrive for work, the staff in the Ops Room has sheets for each of the day's complexes. The complex sheets at that point have labelled rows and columns, but empty cells. Two members of the Ops Room staff (BP and PP) fill in the cells with projected baggage/people transfers. This must be done close to the time of the complex, and is done differently for the two intended user groups, that is, ramp crew and gate agents. Next, the sheet is copied and distributed to the crews assigned to the upcoming complex. The activity at each stage then, is to (a) receive a document partially completed, (b) add information to the document by accessing resources only now available, and (c) copy the document and distribute it to others, or to a file;

3. A *medium* for notating the events of the complex as they happen, in such a way that irregular or problematic sub-events stand out. Once filled in and distributed, the sheet again becomes a template on which BP and PP add highlighting marks during the complex itself. The pattern of cell highlighting (moving diagonally down and to the right) makes missed or delayed transfers apparent as unhighlighted cells left behind the advancing front;

4. An enduring *physical record* for future reference of what happened during this complex. Once the complex is completed, BP and PP file their highlighted versions of the sheets;

5. A *transparent artifact* that stands in for situations out on the ramp and provides a shared object for communication between people during the course of the complex. In the case at hand, we see BP and PP first communicating in person over the sheet and then PP and the gate crew chiefs communicating, this time on the radio, but again with reference to the sheet. In this respect the complex sheet serves implicitly as the kind of "context support system" proposed by the Knowledge and Work Project at

Turku, Finland (Hellman, 1989). It makes available to practitioners the cooperative structure of their work.

Suppose now that we're interested in the ways the complex sheet technology breaks down (or at least is stretched), given the multiple purposes it is meant to serve. One problem that has been mentioned to us by Ops Room staff involves changes that must be made to the sheet after it has been copied and distributed to ramp and gate crews. Because updating outstanding copies is difficult and because the cost of missing one sheet can be high (we were told a story about baggage transferred to the wrong plane for precisely this reason), the Ops Room waits until the last possible moment before the complex to copy the sheets, and makes subsequent changes only with difficulty.

A second problem is illustrated in this incident as well. In this case the complex sheet must be changed to represent a state of affairs unanticipated in its original design. That is, the sheet for one complex must be notated with information about another complex. In this case, PP has done just that and copies have been distributed. BP has trouble interpreting the added notation, however, and inquires further of PP. Together BP and PP arrive at a change to the notation that better communicates PP's intentions. PP then decides to communicate the change to gate crew personnel, in case they encounter a similar problem.

Designing for Practice

One might look at this situation and conclude that the basic complex sheet format should be modified to support notations referring to subsequent complexes. But for every form, procedure, or routine in a workplace, new situations inevitably occur that stretch the form in unanticipated ways. Similarly, one might argue for replacing the complex sheet with a computer-based representation, so that changes instantly propagate to the versions on the various crews' computers. However, computerized forms have their own problems; for example, the loss of ease with which the document can now be transported.

Incidents like these convince us that, given the inevitable trade-offs, effective design involves a co-evolution of artifacts with practice. Where artifacts can be designed by their users, this development goes on over the course of their use. To some extent, however, an initial design must be based on a projection, in one time and place, of work to be done, often by other people than the designer, in another time and place. Moreover, artifacts are frequently (either intentionally or unintentionally) somewhat intractable in their design, allowing little in the way of modification or redesign over the course of their use. In such cases, either the work practice adapts to the demands of the artifact, or the artifact is not used.

Whatever the relation between design and use in time, space, and participants, some amount of projection is a central aspect of design. This suggests to us the importance, in the design of tools, of realistic scenarios of their use. Designers interested in augmenting or replacing current artifacts like complex sheets do well to understand how they work, as well as what their limits are. In addition, those interested in supporting the design of modifiable artifacts do well to understand the everyday processes of modification exemplified in the complex sheet case described above.

Design realism can be achieved, we believe, through new methods for understanding the organization of work practice in detail. Such understandings provide a basis for involving practitioners directly in the process of reflection and design through, for example, collaborative production and review of videotapes and rapid prototyping. In the remainder of this chapter we focus on the methods we use for understanding work practice: using video records as a tool for reflection.

A Perspective on Work Practice and Design

In every setting where we look at technologies in use, we start with the assumption that work practice is fundamentally social. Basically, this is so in that any activity, whether characterized by conflict or by cooperation, relies on a foundation of meaningful, mutually intelligible interaction. Moreover it is the community, rather than the individual, that defines what a given domain of work is and what it means to accomplish it successfully. Finally, every occasion of work, however individualistic it may appear, involves some others, either in the form of co-workers or of recipients.

This basic sociality recommends that wherever we go we look for the human interactions that make up the work and define what counts as competent practice. Whether in a place like an airport, where work is closely coordinated among multiple participants, or in the case of a single user of a stand-alone machine, getting the work done requires some form of collaborative interaction. Even if the relevant others are absent, as in the case of the single user of the machine, we can still learn much from viewing the machine's use as an interaction with the machine itself, and with the communicative artifacts that its designer left behind. As we explain in detail below, this sociality is what makes analyses of human interaction so relevant to technology design.

Design and Use

A general requirement for the development of new technology is successful translation between the conception of the technology held by its designers and the reality of those for whom it is ostensibly designed. This ranges from international development efforts to introduce new technologies to traditional work ways, through industrial efforts to introduce information systems into modern offices (see Suchman & Jordan, 1988). Where technologies are designed at a distance from the situation of their use, as most are, there is an inevitable gap between scenarios of use and users' actual circumstances.

While design scenarios always simplify the projected situation of use, they can be based in more and less adequate understandings of users' practice (see Blomberg & Henderson, 1990). Over the past decade, we have been working to develop methods through which we can come to appreciate the subtleties of work practice in specific settings and to relate those understandings to design.[2] Our research strategy has been to undertake studies of technology and practice in a range of settings; for example, in a U.S. accounting office, rural Mayan communities in Mexico, Western hospitals, computer research laboratories, at photocopy machines, over various remote communications links, and so forth. Through these studies we have begun to apply perspectives from sociology and anthropology to problems in technology design and use. What we see consistently is that the closeness of designers to those who use an artifact (including the possibility that designer and user are one and the same) directly determines the artifact's appropriateness to its situation of use.

The Politics of Artifacts

Anthropologists of technology like Winner (1980), Latour (1986), and Akrich (1987) have pointed to the intrinsically political nature of technical artifacts. "Artifacts have politics," in Winner's phrase, not only in the sense that they embody the ideologies and agendas of their designers, but in the less obvious, more pervasive sense that they "constitute active elements in the organization of the relationships of people to each other and with their environment" (Akrich, 1987, p. 1). Artifacts take their significance from the social world, in other words, at the same time that they mediate our interactions with that world. Winner (1980) offers the example of overpasses in the neighborhood of Jones Beach on Long Island, New York. By

---

2 The "we" here refers to a group of researchers at Xerox Palo Alto Research Center, Palo Alto, CA.

restricting the height of overpasses to be less than that required by public transit buses, the designer Robert Moses effectively restricted access to the beach to those who could afford a private automobile. More specifically, "one consequence was to limit access of racial minorities and low-income groups to Jones Beach, Moses' widely acclaimed public park" (p. 23).

Ethnography and Interaction Analysis

Informed by these perspectives, our work makes use of two related methods for research: ethnography and interaction analysis. Ethnography, the traditional method of social and cultural anthropology, involves the careful study of activities and relations between them in a complex social setting. Such studies require extended participant observation of the internal life of a setting, in order to understand what participants themselves take to be relevant aspects of their activity. Importantly, this may include things that are so familiar to them as to be unremarkable (and therefore missing from their accounts of how they work), although being evident in what they can actually be seen to do.

Interaction analysis is concerned with detailed investigation of the interaction of people with each other and with the material environment.[3] Our use of interaction analysis is inspired by prior work in anthropology and sociology, particularly ethnomethodology and conversation analysis (see for example Atkinson & Heritage, 1984; Garfinkel, 1986; Goodwin, 1981; Goodwin & Heritage, 1990; Heritage, 1984; Sacks, Schegloff, & Jefferson, 1974). In work settings, where our studies have been centered, our analyses focus on the joint definition and accomplishment of the work at hand, through the organization of interaction and the use of supporting technologies and artifacts.

Situations of Use

Our goal in doing studies of work is to identify routine practices, problems, and possibilities for development within a given activity or setting. In the case of the airport operations room, we learned about the difficulties of propagating changes to complex sheets by recording an instance of everyday trouble, as well as by hearing Ops Room members' stories about misdirected baggage. The ideal site for investigations of technology in use, in our view, are these "naturally occurring" occasions of work activity, in the settings in

---

[3] For more details on both interaction analysis and its application to design see Jordan, Henderson, and Tatar (in prep.).

which such activities ordinarily take place. That is, in settings designed by the participants themselves, over time, rather than contrived by the researcher for purposes of the analysis. The first task for such studies is to come to understand peoples' current work practices, using the technologies that are available to them.

A variation on this approach is to investigate the organization of work practice and the use of technologies in situations that we, as researchers, construct. Within those situations, we invite the participants to use whatever tools they choose, and to organize their work in whatever ways they choose. So, for example, we might be interested in understanding how people use shared drawing spaces such as whiteboards or poster paper to do joint design work (Tang, 1989). We ask people who use whiteboards in the course of their everyday practice to come and do some of their own work, but at a time and place that we propose, and with video cameras set up to record their interaction. A further variation occurs when we have an early prototype of a new tool, and invite people to come in to use the prototype, again to do their work but in a time and place that we propose (see for example, Minneman & Bly, 1990).

A situation constructed for purposes of investigation, in contrast to a study carried out wherever the activities of interest are under way, is itself a kind of prototyping. That is, rather than experimental "control," we are really after a way of prototyping either our methods for looking at the details of work practice, or a tool, ideally based on prior study in actual work settings, that we are not quite ready to put out into the world. Like all prototyping, such studies should be done as rapidly as possible, with as strong a feedback mechanism as we can design. And as soon as we can, we should get ourselves out into the world again.

What We Record

In recording within actual work settings, our strategy is to find multiple solutions to the problem of capturing the complex activities that work involves.

Setting-oriented records are made using a stationary camera, positioned to cover as much as possible of the activity in a given physical space. The use of the complex sheet described earlier was recorded with camera and microphone positioned to include all four seated operations room workers in the field of view, and to pick up as much of their talk as possible.

Person-oriented records, in contrast, attempt to understand the work from a particular person's point of view. Depending on the nature of their work, this may require recording equipment that travels with the person being recorded. For example, in order to understand the use of the complex sheet by ramp crew members in more detail, we would need to accompany them as they move around the planes and in the terminal.

An object-oriented record would track a particular technology or artifact. We might trace a complex sheet from the time it is first generated (as an empty matrix) through its markup and highlighting by operations room members, to its filing and possible eventual recovery by auditors. Instrumenting the computers used by operations room personnel to provide machine audit trails yields a different sort of object-oriented record.

Finally, task-oriented records may require multiple recordings of distributed individuals working toward a common goal. A prime example of a task-oriented record would be if during the aircraft swap described above we could have had access not only to activity inside the operations room itself, but also at the two relevant gates and out on the ramp.

How We Work as Researchers

The first analysis we undertake after recording an extended sequence of activity is a rough content log of the entire videotape. Such a log describes observed events and indexes them chronologically by clock time. This rough summary of what happened can expedite the process of searching for particular remembered instances. In addition, the log identifies issues raised in the course of viewing the tape again. Such issue-based logging might include, for example, marking an event as being of inherent interest and possibly representative of a new theme or category (e.g., extending a complex sheet across two complexes); labelling an event as being representative of a given theme or category (e.g., modifiability of artifacts); or generating proposals for further fieldwork (e.g., studying the use of the complex sheet on the ramp).

Frequently, a sequence of recorded activity is picked out as being of particular interest and deserving of more systematic analysis. A first step toward such analysis is a careful transcription (see Jefferson, 1984). Though transcribing talk is almost always central to any interactional study, detailed sequential analyses can also focus on non-vocal dimensions of interaction such as the participants' gaze, gestures, and body positions (see Goodwin, 1981; Heath, 1986; Kendon, 1985). Furthermore, in our work, we have found it valuable to analyze activity in terms of peoples' interactions with various artifacts or technologies. This requires detailed transcriptions of the actions of people using machines, concatenated with machine records to show the interactional organization between person and machine, and the interrelation of that human-machine interaction with the talk and activity of other people (see, for example, Suchman, 1987, chap. 6).

The task of interaction analysis is to uncover the regularity and efficacy of peoples' relations with each other and their use of the resources that their environment affords. To see this regularity requires a productive interaction between an emerging theory and a rich body of specific instances. As we accumulate studies of interaction in specific settings we begin to develop a set of both general observations (e.g., the interorganization of talk and nonvocal activity) and specific analyses (e.g., the specialized concerns and practices of people in this particular domain and/or this particular setting). As an analysis begins to develop over a body of videotapes, it becomes necessary to construct "collections"; that is, instances of interaction that one wants to see as a class. In setting such instances side by side, the commonalities and distinctive features among them are made more visible:

> For example, on one videotape of an American hospital birth it was noted that at the beginning of a particular uterine contraction the eyes of all those present—nurse, husband, medical student—went to the electronic fetal monitor by the woman's bedside. A behavioral pattern was proposed that when a monitor is present, birth attendants' eyes will move to that equipment when a contraction begins. The pattern was checked against other contractions in this particular labor, on tapes of other American hospital births where monitors were present, as well as against monitored contractions in European hospitals. It was found that the pattern held overwhelmingly. In the few cases where it did not, evidence for some competing local activity was available to explain the discrepancy, e.g., the woman was very distraught or the doctor was just at that time doing an examination and therefore had his or her eyes on her. When examining videotapes of births where monitors were not used, it was found that crossculturally in the absence of monitoring equipment, the focus of attention almost always shifts to the woman when a contraction begins. A generalization was then proposed which states that in the presence of high-technology equipment the attentional focus of medical personnel as well as of non-staff attendants shifts from the patient to the machinery. (Jordan, Henderson, & Tatar, in prep.)

Having a videotape in hand in no way eliminates the need for thoughtful interpretation of the meaning of the events it records. However, video-based interaction analysis affords a powerful corrective to our tendency to see in a scene what we expect to see:

> For example, in certain circumstances we expect a couple who are in close physical proximity and who are smiling at each other to

also touch each other. Observers frequently report that they have seen such touches even though on replay it is clear that none occurred. Errors of this sort are invisible on a paper-and-pencil record, but a tape segment can be played over and over again, and questions of what is on the tape versus what observers think they saw can be resolved by recourse to the tape as the final authority. This jolting experience of periodically having one's confidence in what one (thinks one) saw shaken, instils a healthy skepticism about the validity of observations that were made without the possibility to check the record more than once. (Jordan, Henderson, & Tatar, in prep.)

The work of interaction analysis is time consuming and labor intensive for the researcher. It requires the painstaking work of viewing, transcribing and searching videotapes for sequences relevant to a developing analysis. However, a compelling case for the value of this labor is made in Jordan, Henderson, & Tatar (in prep.):

Even for a trained observer, it is simply impossible to note the overlapping activities of several persons with any accuracy or any hope of catching adequate detail. Consider an excerpt from fieldnotes, a paper-and-pencil snapshot of a birth: "midwife bathing baby; mother in hammock; father out to dispose of placenta; grandmother rummaging in cardboard box" (Jordan, 1983). The videotape provides an incomparably richer record. The kind of talk (or silence) going on at the time, the procedural details of the bath, the mother's eyes on the infant, the grandmother's rummaging for oil and baby wraps, the looks, the body orientations, all of that is lost and probably not recoverable from memory. It is also not recoverable from the memory of participants by interviewing after the fact. Similarly, the details of manipulative procedures such as the arranging of flight-status strips on the work panel of air traffic controllers (Anderson, forthcoming) or the details of moving a cursor while text editing, are impossible to capture in words, both because of the density of behavioral details and because we lack a ready descriptive vocabulary for bodily behavior which could be captured in notes.

This work is not easily delegated. Particularly where conceptual categories are under construction, it is difficult to instruct another as to where to look. More importantly, it is precisely in the repeated, careful working through of the primary materials that theoretical insights arise. In this way, analysis is something like iterative design. Articulations of themes and categories arise from familiarity with the materials and are constantly reevaluated against those materials. This in turn renews and extends one's familiarity. Furthermore, the identification of new themes and categories can lead one to return to the field or workplace to gather new materials.

Tools for Activity Analysis

As we have seen, the work of video-based interaction analysis involves a continual interweaving of multiple activities: viewing and re-viewing video records both individually and in groups, generating activity or "content" logs for each record, conducting detailed sequential analyses of selected portions of the records, integrating multiple records (often in different media) of the same activity, identifying conceptual categories and gathering "collections" of instances, and finally, juxtaposing multiple analytic perspectives on the same activity.

Assume for a moment that we have obtained a set of video records of activity in a workplace, including practitioners' work with a particular technology of interest. What tools, in particular what computer-based resources, are available to help manage and manipulate them? The sort of video analysis discussed here suggests possibilities for computer support: the need to generate useful abstractions while retaining their relation to unique instances and the need to capture multiple perspectives on the same piece of recorded activity.

Figure 3 shows part of an ongoing analysis of the operations room activity discussed earlier. The image is taken from the screen of a computer-based "activity representation" tool that we are in the process of developing.[4] Our tool is an example of a hypermedia system in that it represents information in linked, browsable networks.[5] Nodes in the network could include verbal transcripts, records of participants' gaze or other non-vocal activity captured in some graphical notation, brief comments or annotations, or indeed, a videotape itself.

Three "windows" appear in Figure 3: a text window displaying a portion of the verbal transcript, a graphics window depicting our understanding of the three-way aircraft switch, and a three-paned "worksheet" window used to control and annotate the video record. The worksheet's "Content log" pane serves as a canvas on which bits of text and graphics are positioned, ordered chronologically according to the time of the events they depict.

---

[4] The tool described here was designed and prototyped in collaboration with Jeremy Roschelle and Roy Pea of the Institute for Research on Learning, with support from Apple Computer.

[5] For an introduction to hypermedia and a survey of the major systems, see Conklin (1987).

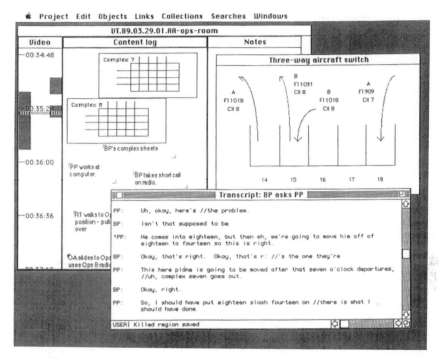

Figure 3. A computer-based environment for video analysis.

The "Video" pane includes ruler-like markings indicating relative time from the start of the tape. The small scroll bar appearing just under "00:35:24" indicates the current position of the videotape. When the videotape is playing this scroll bar moves down the pane. The three filled rectangles in the video pane mark the locations of *active links*. As the scroll bar moves past these link markers, the linked materials are automatically displayed in separate windows. The larger rectangle is linked to the graphics displayed in the "Three-way aircraft switch" window while the two smaller rectangles link to positions within the transcript.

The tool enables us to maintain and interrelate multiple perspectives on a work activity. Perspectives can differ on the basis of the dimension of activity picked out (e.g., talk, gaze, gesture, use of technology), or the viewer (e.g., different researchers, designers, and practitioners). In doing cooperative design we need to share materials: If video provides the catalyst for multidisciplinary collaborative work, then tools with which to study, represent, and discuss video records provide the ongoing support.

Who Participates

Most video-based analysis of activity makes use of a preconceived coding scheme, carried along largely unchanged from one study to the next. Our approach, in contrast, is to craft and refine our conceptual framework through the ongoing work of analysis. A basic strategy is to use multiple perspectives, drawing on designers, people informed about interaction analysis, and those able to unravel the intricacies of practice in a given domain. Although our use of these methods to date has been primarily in the context of traditional research, we believe they can be developed in the service of co-operative design. In our current working group, researchers bring tapes and transcripts from their respective projects to a weekly two-hour meeting.[6] The group offers a forum in which researchers can identify the possibilities that a particular tape affords. The entire working session is audiotaped, so that with those proposals and insights in hand, individual researchers or project groups can direct their more intensive work outside the lab to a particular line of analysis. Lines of analysis developed outside the lab in turn can be brought back to the larger group for reactions and new insights.

In a typical session, someone provides a brief introduction to the setting of the recorded activity and the project's interests in coming to understand that activity better. The group then works from a running videotape that is stopped whenever anyone observes something of interest. Group members propose interpretations for what is going on, attempting to provide some grounding or evidence in the materials at hand. Such proposals usually require repeated reviewings of the tape, until alternative formulations are laid out or the group agrees on one interpretation. Often an observation leads to further questions to be asked out in the field, at which point the questions are noted and viewing continues.

As well as being the focus of discussion for researchers and designers, video records can afford a basis for conversation with work practitioners. In a process developed most extensively by Frankel (1983) in working with student physicians, participants in a recorded interaction review the tape along with researchers. They are invited to stop the tape at any point and offer comments on what they see. Their comments are audiotaped, and then flagged at the place on the videotape that occasioned them.

This gives some idea of how participants structure the event, i.e. where they see significant segments as beginning and ending; it

---

6 We do this regularly through the Interaction Analysis Laboratory, a joint activity of Xerox PARC and the Institute for Research on Learning in Palo Alto, and through the Design Interaction Analysis Laboratory at Xerox PARC.

also gives information on troubles that may be invisible to the analyst, on resources and methods used by participants to solve their problems, and on many other issues of importance to participants. Frankel, for example, asked patients and physicians who had been videotaped during medical consultations to stop the tape when they saw something of interest. There was a substantial overlap in *where* they stopped the tape, but the explanations they gave for *why* they stopped the tape were widely divergent between patients and physicians, indicating substantially different views of what their interactions were about. Elicitation based on video tapes has the advantage of staying much closer to the actual events than if one were to ask questions in a situation removed from the activity of interest to the researcher. (Jordan, Henderson, & Tatar, in prep.)

With these experiences as background, we hope to engage practitioners themselves in the process of analysis in order to collaborate on both work and system development.

Applying Video Analysis to System Design

Together with colleagues at Xerox Palo Alto Research Center (PARC), we have been exploring the technological as well as the social-interactional resources of collaborative work practice, focusing on the use of available artifacts. In particular, this has involved extensive studies of the use of shared drawing surfaces (see, for example, Tang, 1989; Tang & Minneman, 1990). To illustrate the power of video-based study for design, we trace the experience of two of our colleagues building a prototype shared drawing tool.

The objective of these studies is to understand how the use of the shared workspace both supports and is organized by the emergent structure of the ongoing activity. The conventional view of a shared writing/drawing space as a medium for the communication and recording of individual ideas is displaced by a detailed analysis of the use of such a shared workspace. By looking closely at each instance where participants in a working session put pen to paper, we begin to understand the complex sequential relationship between writing and talk in the joint accomplishment of the work at hand.

Scott Minneman and Sara Bly are currently engaged in a process of design that incorporates these observations. In what follows we briefly summarize their experience; the interested reader should consult Minneman and Bly (1990) for a full discussion.

Prior to designing Commune, their multi-user drawing tool, Minneman and Bly were involved in studies of collaborative drawing activity using traditional media. In particular, Bly conducted studies of a pair of user interface designers working in three communicative environments: face-to-face, in offices joined by a slow-scan video link, and over the telephone. This research revealed the work required to maintain coordination and co-presence as interactional and representational resources became less accessible and more difficult to share. In the Office Design Project (Weber & Minneman, 1987; Stults, 1988), three architects collaborated for two days on the design of an office space, communicating only via audio and video links. As Minneman and Bly (1990) put it:

> The results of these examples of using Media Spaces for everyday drawing activities indicated that better tools were needed to support individuals and groups collaborating remotely. It was also believed that such tools might easily generalize to support collaborative design work in other environments. The need for finding a solution to shared drawing support was apparent: the next step was to understand how and why drawing surfaces were already being used. (p. 2)

One resource for gaining such understandings is a workgroup at PARC consisting of anthropologists, computer scientists and designers called the Designer Interaction Analysis Lab (DIAL) (described in Tatar, 1989). This group meets regularly to apply the methods of interaction analysis to concrete problems in system design. The group views videotapes of people working with traditional as well as prototype technologies, noting the subtleties of established work practices with familiar artifacts, as well as the problems and adaptations encountered by practitioners using new technologies.

> One major finding from [DIAL-related] studies has been the importance of the *process* of the collaboration in contrast to the actual artifacts or markings on the shared work surface. Process includes the use of gestures, the ability to interact in the same space, the use of references, the lack of distraction caused by the drawing action, and the feeling of close interaction. Observing the use of drawing surfaces has pointed out the potential value of providing not only shared marks in a tool for distributed use but also support for the interactive process. (Minneman & Bly, 1990, p. 3)

The studies left several questions unanswered, however, suggesting "that the capabilities of a shared drawing tool might best be *evolved* through use, rather than through analysis and design alone" (p. 3). This led to a minimalist design for the Commune shared drawing tool, modeling the user interface after a single shared sheet of paper with marking pens. The first prototype could thus be developed in short order and made available to casual "walk-up" users almost immediately thereafter. Videotapes were made of these encounters as well as of one extended working session by two user interface designers. Initial study of this data indicates the "ease with which almost all users began using the system..." (p. 5).

As Minneman and Bly (1990) point out, such careful use-based prototyping isn't always the most natural path for designers:

> Despite the initial success of our development cycle for Commune, there is considerable temptation to revert to design evolution based solely on intuitions and the technological possibilities offered by the prototyping platform. (p. 6)

Our purpose in applying video analysis to design is to resist such temptation by developing methods that encourage the move from analysis to development to further analysis. The goal is to tie both intuitions and technological possibilities to detailed understandings of real work practice.

Reflection and Design

During meetings in the course of assembling this book, Pelle Ehn provided what we found to be a helpful formulation of the relations among the book's authors. Some of us are reflective practitioners, he said, referring to the book by Donald Schön (1983). And others of us are practicing reflectors. What we share is the goal of bringing the two together. Theorizing or reflecting and engaging in some form of practice are essential to every kind of human endeavor, be it research, design, or any other form of work practice. To develop simultaneously our ways of working together and our technological systems will require a joint enterprise that recognizes, encourages, and develops both the quality of our work practice and our powers of reflection, with respect for both and without privilege to either.

In Figure 4, this joint enterprise is depicted in the form of a triangle linking three perspectives: research, design, and practice. We deliberately have not said researchers, designers, and practitioners, because our goal is to view these not only as a division of labor between participants in the system development process, but as places from which to look. Depending on where one stands, at which corner of the triangle, one adopts a different perspective.

As a consequence of personal preference, experience, and training, we are likely to distribute ourselves differently across these perspectives. We believe that the most satisfactory organization may take the form of collaborations wherein some of us take on primary responsibility for one of these perspectives, and others of us for other perspectives. However, we want to think of these, not as absolute, fixed positions but as relative to each other. And we want all to have facility at moving between perspectives.

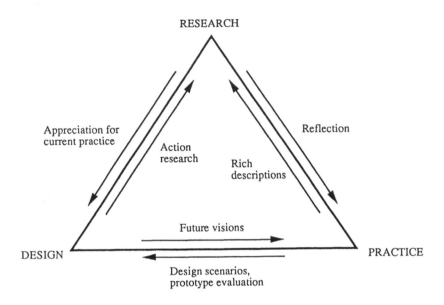

Figure 4. Video as a medium for conversation and learning.

In Figure 4, the arrows appearing along the edges of the triangle indicate the kind of contributions that one perspective can make to another:

- Using the perspective of research, practitioners reflect on their current practice;
- Using the perspective of design, practitioners envision future work practices and new technologies;
- Using the perspective of practice, researchers gain the insights that only confrontations with reality can provide;
- Using the perspective of design, researchers bring an action-oriented involvement to their traditional analytic work practice;
- Using the perspective of practice, designers working with users better understand the implications of prototypes and scenarios for new designs;
- Using the perspective of research, designers gain new insights from observing current practice so as to respect the present when designing for the future.

Examples of such pairwise collaborations can be found throughout research and industry wherever people concern themselves with work practice and technology. What we now hope to foster are collaborations among all three perspectives simultaneously, that is, activity occurring in the center of the triangle. Consider the case of the videotaped working session excerpted at the start of this paper. In a joint session of practitioners, researchers and designers looking at the recording, the various groups might have overlapping but not identical goals. All might share the goal of understanding better what happened in the interaction, though the participants might be reflecting on their work together, the researchers investigating the interactional organization of the work, and the designers searching for ideas for new prototypes. We hope that such cross-perspectival cooperations become the norm in the future, particularly in cases that today involve only one of the parties. We believe that the video medium can catalyze such collaborations to the greater understanding and imagination of all concerned.

Acknowledgments

We are grateful to Liam Bannon, Susanne Bødker, Bengt Brattgård, Bo Dahlbom, Pelle Ehn, Joan Greenbaum, Austin Henderson, Berit Holmqvist, Jana Kana-Essig, Morten Kyng, Jens Kaasbøl, and Jesper Strandgaard Pedersen, for comments on earlier drafts of this paper.

References

Akrich, M. (1987). How can technical objects be described? Paper presented at the Second Workshop on Social and Historical Studies of Technology, Twente University, The Netherlands.

Anderson, R. J., Hughes, J. A., Schapiro, D. Z., & Sharrock, W. W. with Harper, R. & Gibbons, S. (forthcoming). Flying planes can be dangerous: A framework for analysing the work of air traffic control. Submitted to *Human-Computer Interaction*. Hillsdale, NJ: Lawrence Erlbaum Associates.

Atkinson, J. M. & Heritage, J. C. (1984). *Structures of social action: Studies in conversation analysis*. Cambridge, UK: Cambridge University Press.

Blomberg, J. & Henderson, A. (1990). Reflections on participatory design: Lessons from the Trillium experience. *Proceedings*

CHI' 90 Human Factors in Computing Systems (Seattle, April 1-5). ACM, New York, 353-360.

Conklin, J. (1987). Hypertext: A survey and introduction. *IEEE Computer, 20* (9), 17-41.

Frankel, R. (1983). The laying on of hands: Aspects of the organization of gaze, touch and talk in a medical encounter. In S. Fisher & A. Todd (Eds.), *The social organization of doctor-patient communication* (pp. 10-33). Washington, DC: Center for Applied Linguistics.

Garfinkel, H. (Ed.). (1986). *Ethnomethodological studies of work.* London: Routledge & Kegan Paul.

Goodwin, C. (1981). *Conversational organization: Interaction between speakers and hearers.* New York: Academic Press.

Goodwin, C. & Heritage, J. (1990). Conversation Analysis. Prepared for *Annual Review of Anthropology,* 1990.

Heath, C. (1986). *Body movement and speech in medical interaction.* Cambridge, UK: Cambridge University Press.

Hellman, R. (1989). Emancipation of and by computer-supported cooperative work. In *Scandinavian Journal of Information Systems, 1,* 143-161.

Heritage, J. C. (1984). *Garfinkel and ethnomethodology.* Cambridge, UK: Polity Press.

Jefferson, G. (1984). Caricature versus detail: On capturing the particulars of pronunciation in transcripts of conversational data. *Tilburg Papers on Language and Literature No. 31,* University of Tilburg, Netherlands.

Jordan, B. (1983). *Birth in four cultures* (3rd ed.). Montreal: Eden Press.

Jordan, B. (1987). High Technology: The Case of Obstetrics. *World Health Forum 8*(3), 312-319. Geneva: World Health Organization.

Jordan, B., Henderson, A., & Tatar, D. (In preparation). *Interaction analysis: Foundations and practice.* Palo Alto: Xerox Palo Alto Research Center.

Kendon, A. (1985). Behavioural foundations for the process of frame attunement in face-to-face interaction. In G. Ginsburt, M. Brenner, & M. von Cranach (Eds.), *Discovery Strategies in the Psychology of Action.* London, UK: Academic Press, 229-253.

Latour, B. (1986). *How to write* The Prince *for machines as well as for machinations.* Working paper, Seminar on Technology and Social Change, Edinburgh.

Minneman, S. & Bly, S. (1990). Experiences in the development of a multi-user drawing tool. *Proceedings of the Third Guelph Symposium on Computer-mediated Communication.* Ontario: University of Guelph.

Sacks, H., Schegloff, E. A., & Jefferson, G. (1974). A simplest systematics for the organization of turn-taking for conversation. *Language, 50,* 696-735.

Schön, D. (1983). *The reflective practitioner.* New York: Basic Books.

Stults, R. (1988). Experimental uses of video to support design. Palo Alto, CA: Xerox Palo Alto Research Center.

Suchman, L. (1987). *Plans and situated actions.* Cambridge, UK: Cambridge University Press.

Suchman, L. & Jordan, B. (1989). Computerization and women's knowledge. In K. Tijdens, M. Jennings, I. Wagner, & M. Weggelaar (Eds.), *Women, work and computerization: Forming new alliances* (pp. 153-160). Amsterdam: North-Holland.

Tang, J. C. (1989). *Listing, drawing and gesturing in design: A study of the use of shared workspaces by design teams.* Technical report SSL-89-3 (Ph.D. Dissertation, Stanford University). Palo Alto, CA: Xerox Palo Alto Research Center.

Tang, J. C. & Minneman, S. (1990). VideoDraw: A video interface for collaborative drawing. *Proceedings CHI'90 Human Factors in Computing Systems, Seattle, April 1-5* (pp. 313-320). New York: ACM.

Tatar, D. (1989). A preliminary report on using video to shape the design of a new technology. *SIGCHI Bulletin, 21,* 2, 108-111.

Weber, K. & Minneman, S. (1987). *The office design video project* [Videotape]. Palo Alto, CA: Xerox Palo Alto Research Center.

Winner, L. (1980). Do artifacts have politics? *Daedalus, 109,* 121-136.

5

Language, Perspectives and Design

Berit Holmqvist and Peter Bøgh Andersen

The inversion of viewpoints occasioned by Structured Analysis is that we now present the workings of a system as seen by the data, not as seen by the data processors. The advantage of this approach is that the data sees the big picture, while the various people and machines and organizations that work on the data see only a portion of what happens. As you go about doing Structured Analysis, you will find yourself more and more frequently attaching yourself to the data and following it through the operation. I think of this as interviewing "the data."

DeMarco, 1978, p. 49

In opposition to the dominant data flow perspective, expressed in this quotation from Tom DeMarco, this chapter presents a positive understanding of users as sources of information. This does not mean that the data-flow perspective is rejected, but if this perspective is the only one adopted, and if the working of a system is not also "as seen by the data processors," then system development is going to miss crucial information.

This chapter does not present the users' point of view but rather the activities of the workers and the working of the system as seen from a linguistic point of view. We introduce linguistic concepts and methods for analyzing interpretations of work and organizations. The concepts that are part of the stock in trade of traditional structural linguistics include *sign, code, perspective, semantic field, syntax,* and *case,* and are successively introduced and explained, so that the reader ultimately should be able to understand system development from a linguist's point of view.[1]

We view computers as controlled by means of signs, either verbal or pictorial. Like other sign-bearing media such as books, newspapers, film, or video, they should present relevant and understand-

---

[1] A more comprehensive account of the theoretical framework can be found in Bøgh Andersen (1990).

able information. This is not always the case, as illustrated in the following taperecording, where a worker tries to explain the meaning of the log in the system she uses.[2]

L: They are different files, it is how much *frags* there are, how many errors there are, *hots*, what was it, it was ... do you remember?

X: Yeah.

L: I don't remember shit of what I learned there.

The log is clearly incomprehensible to the worker. It is full of "computerese," and at first sight one is tempted to blame the disfunctionality on that. But as linguists we knew that foreign words could be learned if they occupy a relevant place in a language system. The essential point is that computers present a particular interpretation of the tasks performed on them. If that interpretation is irrelevant to the workers actually performing the task, it prohibits the understanding of the system. In the following quotation, two workers are discussing the same system, the topic now being the control area of the screen where descriptions of the tasks are displayed.

H: Like, if we say, now it says "completion" up here, but then, when I finish the pile, then it says "production," how come?

R: Because then you begin a new "production," then, when you type, when you have typed the first one and the last one, then it says "completion."

H: Yes.

R: It always says here exactly what you're doing.

In this case, H has a different idea of "what she is doing" than the system has. Will H maintain her own interpretation, or will she come to accept that the system in fact "says exactly what she's doing"? In the latter case, whose interpretation of her work is it she adopts when accepting the interpretation given by the system?

To construct computer systems is also to construct and communicate an interpretation of the tasks and the organization in which the system is used. Although we do not believe that a system designer should design the computer-based sign system as a copy of the ex-

---

2 The examples are taken from an empirical research project, *Professional Languages in Change*, conducted by the authors in 1986 at the Postal Giro Office in Stockholm, Sweden. The aim of this project was to describe how a professional language changes when computers are introduced at work. The examples are in most cases translated from Swedish to English, which means that we don't get a "true" but rather a "close" picture of reality.

isting work language, we feel that some knowledge of the latter is a good basis for understanding the part of design that has to do with the users' interpretation of the system. It will give the designer a keener awareness of the design choices in this area. Therefore, we also show how these methods can be included among the techniques for designing computer systems.

Perspectives Differ With Organizational Roles

In a traditional hierarchical organization, it is possible to distinguish between two fundamental roles in relation to the work: one role that performs the work, and another that describes and organizes the work. Linguistically, this means that one either talks *in* the work or *about* the work. In such an organization, there is often a clear distinction between these two. The roles are, of course, not fixed, and their manifestations vary according to the situation. The following is based, however, on the hypothesis that there is a fundamental, visible differentiation of manifestations, and that there is a methodological point in describing people's linguistic manifestations as a product of their varying associations to the work process.

Describers and Performers

In the following example, (1) has been taken from a recorded interview with a coordinator, (2) from a recorded interview with an instructor who teaches the use of the system at office level, and (3) from a recording of the work language at office level. In (1), the interview takes place in the office of the interviewed, in (2), at a table in the bookkeeping department, and in (3), at a work station.

(1) The coordinator:

One of the purposes of the Optical Character Recognition is of course that one can read everything optically—directly, that is—for example, our pay-in slips "C"; because they have complete rows of optic codes, so they never have to pass the B 25.

(2) The instructor:

They code a lot of C-slips that have this optic area—the machines are able to read it, and so they read the slip and we don't have to touch it—when we put it into the frame, the machine reads it and then we lift the blue box, and then they are ready.

(3) The work situation:

E: Well, I think you should do like this: you take a small piece of paper, and then you attach it to the C-slip to tell that it is

> a C-slip. Later, when you have got it back, then you put everything in this box.

X: Well, I'll have to leave this till tomorrow morning, then I'll give it to her, because I had one today that I left for her—a slip that was missing something that she had cut out—we'll take that too, and then I'll put back this C-slip when she ... well

E: Because she is bound to see that one too if you leave it for her, she wants to see the C-slip, too.

In the first two examples, the interviews take place outside the work situation and the persons are talking *about* work. The difference is that the person in (1) does not normally take part in the work process in question, whereas the person in (2) does. In (3), the persons interact with and comment on the process *in* which they both take part.

When we look at reality, we always concern ourselves with parts. We are interested in one slice of the cake, and we reject the rest. In simple terms, what we are interested in is usually something we want to learn about, or think that the person we are talking to wants to learn about, and what we reject are things we do not need to know, or which we already know. This has direct consequences for how we select things to talk about.

In this case the selection has been done by us; we have picked out bits from long conversations. Furthermore, in the first two cases, the participants' actual selection is determined by the questions asked by the interviewers, the subject being "advantages of the new system"; and in the last case, it is determined by the working process—the persons talk, in a simple manner, about problems as they appear in the course of the work.

To Be General or Specific

In both (1) and (2), the attitudes toward the work process are of a general nature. Both persons are able to describe the general premises of the process without including specific persons or objects of time and space. This is demonstrated by their choice of indefinite pronouns, for example:

one can read
a lot of C-slips

and in the use of nominalizations:

the Optical Character Recognition

and temporal/causal relations combined with timeless present tense:

because they have .... *so* they never have to pass

Moving on to (3), we will see that this example is not in the same vein as the others, although it deals with the same working process. The description of the working process is not of a general nature; instead, the language is used for interacting with work and commenting on it. This means that the statements are related to time and space. They talk about specific work objects and specific persons that can be pointed out in the room. This is shown in the use of the definite article:

> then I'll put back *this* C-slip

and personal pronouns:

> then *you* attach

The actions they talk about take place within a certain period of time:

> leave this *till tomorrow morning*

and in a certain sequence of time:

> take a small piece of paper, *and then ...*

When they talk about work they adopt a perspective of generality, whereas talking in work implies a perspective of specificity.

To Be Normative or Descriptive

The staff is working at work station B 25. In the first two examples, we find statements about the flow of data and the handling of material that are directly contradicted in the third situation. With regard to the C-slips (paying-in forms that are fully pre-coded), it is said that

> they never have to be passed on to B 25

and that

> we don't have to touch them any more

but in (3) there are indications that the C-slips are in fact included in the work at the work station:

> you take a small piece of paper, and then you attach it to the C-slip
> to tell that it is a C-slip
> then I'll put back this C-slip

This does not mean that (1) and (2) are lying, it simply means that they have both decided not to talk about the concrete reality; instead, they talk about the system as principle and possibility which can be seen in their use of words for purposes and possibilities:

one of the *purposes* of the Optical Character Recognition
we don't *have to* touch it
the machines *are able* to read it.

While the coordinator and the instructor are normative, talking about
how the work process "ought to" be, the staff is descriptive, talking
about breakdowns in the routine and how the working process "is."
The work process "is" such that the C-slips are often *not* recognized
by the machine, and have to be manually registered, but it "ought
not to be" this way.

When talking about work it is easy to become normative since the
work situation is not present to verify your description. This could
be compared to the top down view of traditional system development. When talking in work on the other hand, deviations from the
norm often become the main conversational subject. Therefore, re-
vealing a normative perspective presupposes a thorough knowledge
about the working situation, including the language that goes with it.

Participants and Spectators

Although the coordinator and the instructor both adopt an "about"
perspective there are differences. These differences are due to their
roles as *spectator* versus *participant*. The coordinator's job is to
establish routines for production and cooperation on an overall level.
Therefore he has to adopt what is usually called a bird's eye perspective. The instructor's job is to teach the routines in direct connection to the work situation. She does not perform the work but
still she is present in time and space, giving a helping hand and good
advice. Therefore she tends to adopt what we will call an ant's perspective.

Talking about Activities or Processes

The linguistic item that reflects their difference in perspective most
clearly is probably the semantic roles appearing in the examples. A
semantic role is a relationship between a verb and another con-
stituent of a sentence describing an event. It is characteristic of the
common semantics of verbs that they have certain roles attached to
them. For instance, a verb like "read" implies that there are readers,
agents, and that they read something, an *object*. They may use a
magnifying glass, an *instrument*, to help them read: "Hanna read the
book through a magnifying glass."

It also implies that the reading is done in *time* and *space*: "at the
kitchen table yesterday." Sometimes there is also a *cause* built in:
"Hanna who has bad eyes read.. ." The semantic role is supposed
to be part of the linguistic deep structure. The linguistic deep struc-

ture is the abstract system behind our concrete usage of language, which in turn is called the surface structure, according to the theory of transformational grammar. The semantic roles can manifest themselves as different grammatical roles (sentence members) in the linguistic surface structure (see Fillmore, 1968).

Now let us look at how the coordinator and the instructor expressed themselves in the example given. The following constructed sentence represents the simplified deep structure:

Workers can read slips   by the means of an OCR

agent   *object*   *instrument*

It is now possible to vary the grammatical manifestations of the semantic roles on the basis of what one wants to emphasize.

The coordinator expresses himself like this:

everything one can read optically

Neither the agent nor the instrument nor the object is made the grammatical subject; instead a "subject dummy" has been introduced in the form of the indefinite pronoun *one*. And in the next one

(One of the purposes of) Optical Character Recognition

the event has been turned into a nominalization, that is a thing that can be given a grammatical role in a new sentence. The instructor, however, behaves a bit differently:

the machine reads it

and makes the instrument the grammatical subject in all three sentences.

The coordinator's perspective is the perspective of a flow, in which the focus is on the paper and information (*everything that one can read*), and the actions on them viewed as objects (*the optical recognition*). This is not incidental, but fully in line with the traditional data flow perspective within system development, as DeMarco expressed it in the introduction.

The instructor focuses on machine activities that make paper and information flow. But this does not mean that she only sees a portion of what happens; she just sees more detailed portions. For example, the procedure which the coordinator describes as *the optical recognition* is described in three stages by the instructor:

we put it into the frame
the machine reads it
we lift the blue box

Whereas the coordinator rejects people as well as activities, the instructor includes the physical activities before and after the technical processing, with persons as agents. Whereas the coordinator goes on to talk about the work station, through which the "flow" does not pass:

so they never have to pass the B 25

making the slips the subject of his sentences, the instructor stops at the workers, making them subjects and the slips objects:

we never have to touch them

The coordinator's job is to establish routines for production and co-operation on an overall level. His work object is *processes* by means of which information is transmitted. To him, the information plays the main part in the game. The instructor's job is to teach the routines, and her work objects are *workers performing concrete activities*. To her the workers or their stand-ins performing the activities are playing the main part.

Semantic Fields in Work Language and System

In the previous section, we contrasted different fragments of the language usage in the organization in order to illustrate the difference in perspective when persons with different functions describe the same phenomenon. In this section we will contrast system design with the work language around it. Our example will be the PGP-system introduced at the Postal Giro around 1986.

The terms of the PGP-system for the *work material* are partly taken from the old work language in the bookkeeping department at the Postal Giro. But when the system was built, new terms were introduced; the old terms had to make room for new ones and were therefore given a new content.

The words for the *tasks* were taken from other sublanguages. In spite of the fact that the content of the new words is almost the same as the content of the words in the old work language, the words for tasks were rejected by the workers and were replaced by other innovations.

To describe this we use the concept of semantic fields. The concept is traditionally used within linguistics to make comparisons among the ways national languages classify phenomena in the world. As an illustration we give the following example of the extension of similar words in Danish, Swedish, and English.

Danish Swedish English

skov	skog	forest
træ	trä	wood
	träd	tree

Figure 1. A model of a semantic field.

However, instead of comparing different national languages we just compare different sublanguages within a company.

Old Words in a New System

First we are going to compare words for work material in the old work language, in the system, and in the new work language.

In the construction of the system at the Postal Giro, linguistic terms for the work material from the old work language have been applied as far as possible, but new words are also introduced. This leads to changes in the semantic fields.

As shown in Figure 2, the term "job" has been introduced into the system; it is taken from the work language of computer professionals, where a "job" is simply a term for a certain amount of data, which the machine processes as a whole. Thus, "job" has come to occupy the place previously possessed by "bunt" (pile), (i.e., an amount of working material which is relatively easy to handle, but also arbitrary), and this causes "bunt" to be pushed into the location of the old word "BO" (i.e., the current transactions of one client ; a certain number of slips together with a payment order signed by the client). This chain reaction sets the meaning of the word "BO" in motion, so that it becomes more limited and now refers to the payment order only. (The corresponding concept is lacking in the old work language, as indicated by the large shaded area.)

This process is copied in the new work language with one exception: the work language version of "job" and "bunt" refers only to the physical cards that are actually in the hands of the workers, while the extension of the words in the computer system is "all slips read by the OCR together with the manually registered ones." This restriction of extension is symbolized by the shaded areas in Figure 2.

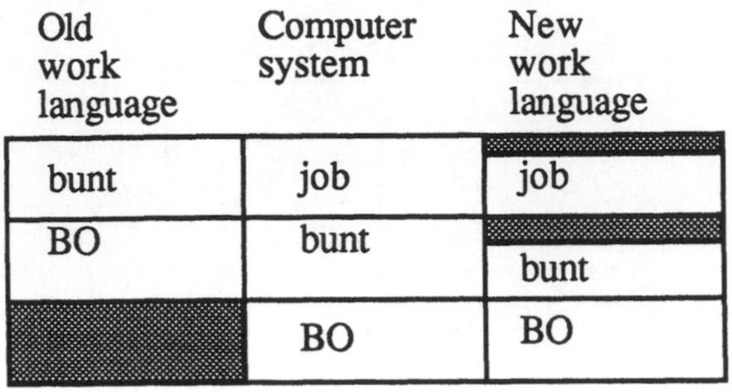

Old work language	Computer system	New work language
bunt	job	job
BO	bunt	
		bunt
	BO	BO

Figure 2. A semantic field for work material.

The old work language is based on a perception of the work as handling papers and documents, and an imitation of this has been attempted in the computer system. At the same time, the staff has demanded that they keep the ordinary slips in order "to keep at least a part of reality."

The result is that in the bookkeeping department there are now slightly different meanings of these words. One for the old system, one for (the paper slips of) the computer-based system, and one for (the electronic slips of) the computer system. The old system will progressively disappear; the idea behind an electronic system is to gradually render paper superfluous, but the fact remains that the terms are the same for ordinary slips and electronic signals. Our observations, however, indicate that it has been easy for the staff to adopt this linguistic change. We have not noticed any misunderstandings or conflicts as a result of it. For us as observers it was extremely difficult to know if the staff was talking about paper-based or electronic work objects, but for them as performers in time and space the situation resolved any ambiguities.

New Words in a New System

We have already seen how words from the old work language have been borrowed and readjusted to the new system. In spite of a rather complicated change in the semantic structure, the new language usage is easily accepted by the users. Now we will look at a case where new words have been introduced.

One of the consequences of the electronic system is that several functions that used to be clearly separated in time and space can now be combined and carried out by *one* person at *one* work station without being separated in time. The old work system has two functions, "coding" and "correction," that are carried out in separate units and with the staff changing places every half hour; and one function that is carried out in a separate department, and that, in the work language, is called "the Correction." The system designers have introduced a more complicated and hierarchical description of the work organization, but have preserved the old interpretation of different functions in it, just with new wording. The monitor display contains the overall term "activity," divided into four functions on the next level, "production" being one of them and "other competence" another one. These two resemble the old division of labor with routine tasks performed in the department and specialist tasks performed in "the Correction" department.

"Production" has two subordinate functions on the same level: "completion," which equals the old task "coding," and "balance," which equals the old task "correction."

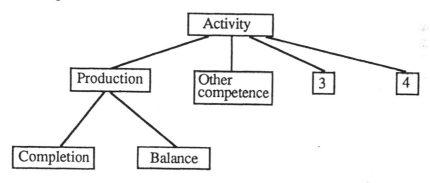

Figure 3. A hierarchical division of tasks.

What is interesting about this is that the functions that used to be clearly separated—spatially, technically, and with regard to roles—and that the staff used to view as separate, when combined in the new system are viewed by the staff as a single function. In their new work language it is called "registering"; the employees neglect the outdated system description of the work organization and construct a new semantic field for tasks which again is based on time and space. The choice is between working at the terminal "registering" or at the OCR, now called "running" or "standing." The following figure, a rough generalization based on the most frequently used "functions" in connection with the work languages of the old and the new systems, illustrates this.[3]

---

[3] The reason why the data recording system has no term (the shaded area) corresponding to the work language term "running," is that it "knows" nothing about the system that controls the OCR-reader, which by the way employs a completely different set of concepts and is in English, not Swedish.

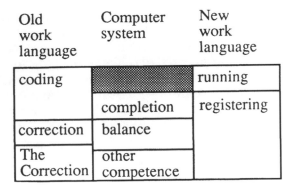

Figure 4. A semantic field for work activities.

Of course, there are technical and historical reasons for introducing this new division of functions into the system, but to the staff, the differences are hardly evident. As far as time and space are concerned, they are in the same work situation when working at the work station, and the terminology is by no means self-explanatory. If the purpose of this terminology has been to create a specific interpretation of the work process, the attempt has not been very successful, as was illustrated in the conversation in the beginning of this chapter.

"It always says here exactly what you're doing," the instructor said there. Does the woman really not know "what she is doing," or does she merely have another perception of it?

The examples in Figure 4 show that the new terms are taken from somewhere other than the work language of the bookkeeping staff. The words "production," "completion," and "balance" do not exist in their old work language; they are taken from other work languages or professional languages. "Balance" is indeed a bookkeeping term, but it belongs in a theoretical situation. It is an educational term. Indeed, it was the main function of "correction" in the old system, but there was no word for this. "Production" is a term from economics. The word "completion" can almost be characterized as a common language term.

But as we mentioned in the introduction this is not a main point. The general meaning of the words are not that difficult to understand. Rather, the point is that the words that are chosen reflect a specific view of the work. The word "completion" is particularly interesting. It reflects the fact that the staff only performs parts of the work. They only add something that has been lost in the Optical Reading Recognition process. It goes perfectly well with the coordinator's flow perspective.

It is not unreasonable to expect that the workers should adopt the word. In fact the greatest part of their daily work, and their main task, is to perform this function. But as seen in the following conversation between the researcher (R) and the worker (H), the worker, in spite of the fact that she is obviously very well aware of the meaning of this specific word, is still not willing to adopt it in her work language:

R: These words up here, "balance" and "completion," did you use them before or are they new words?

H: No, they're new words.

R: Do you use them yourselves?

H: Completion, because when we were coding, we were holding all the slips, even the C-slips.

R: Do you say completion?

H: Yes, because here we only get certain things—if the C-slip is wrong, and then, when there's an error in the ID-section, and maybe it hasn't been able to read one or two digits out of thirty-four, then I simply type these digits he hasn't been able to read, and then I have to complete.

R: You say complete, do you?

H: Well, but we use to say—*anyway, we call this registering.*

R: Did you also use that word before? Did you say registering before? Or did you say coding?

H: No, comple—coding.

R: Registering is also new then?

H: Yes, that is also new—it was separated, because when they were sitting at "the checks" over there [i.e., a special section in the department, where they have had computer systems for a while], then it was registering, it was a little more posh.

"Registering," which the staff prefers, is an overall term for different types of data collection. And yet, here the term is related to data entry in computers. The reason for the adoption of the word is probably (as the example indicates) that the word is familiar in the department. It has been borrowed from a special section that was not occupied with coding and punching, and therefore needed a specific word for their data collection to mark the difference. Now when computers are introduced it can be used in the whole department as something different from the old techniques.

Two different perspectives meet in the system. One describes work as manipulation of physical objects. This is reflected in the reuse of old words for working material. The other one describes work as information processing, reflected in the introduction of the new abstract words for work organization. The staff chooses to accept the first perspective (as seen in their adoption of the new content for old words) and to reject the second (as seen in their refusal to adopt the new words and instead introduce one of their own).

Semantic Fields as a Basis for Designing the Objects of the System

In the previous sections we have given examples of divergence between system and work language; a natural question to ask is: How can we use knowledge about the work language in design, and how might the system have looked like, had the work language been used as a basis for design? In this section, we show examples of how semantic fields can be used to get ideas for design.

The first thing to note is that there is no question of copying or translating work language into computer-based signs. There are two reasons for this. First, the computer is a medium other than the verbal language, and this alone prohibits direct copying. Second, system development normally also means organizational change—changes of task structure, of division of labor, of responsibilities and rights, and of values and norms.

The idea is to use work language as the point of departure for inventing new computer-based signs. It is more like making a film version of a novel than translating a book from one language to another.

The second thing to note is that work language can never be used on an isolated basis. However, after having studied and analyzed it, we can use it to build prototypes that can be used as a source for constructive discussions with users.

Our point of departure will be the previous observation that the semantic fields of the system are different from those of the work language, not only on a superficial level, but in the distinctive features that are used to structure them. Many signs in the actual system are related by part-whole relations, but these meanings are not operative in the work language.

We have seen that the workers used space and time as principles for naming their work processes; this same principle can be observed in many other places in their language. The following quotations are from a speaking aloud session. We asked a worker to tell us what she did while she worked. In her speech, work objects can *go there, come, come forth*;

you know, everything goes there
this one is no good because then all cards come

and she presents her own actions as a journey in a landscape of
cards and tasks, describing the surroundings from her present location: *come to, go to, go back to, be, sit*;

because I came to the last card on this one
when I sit in "comp" then it is "completion"

Her own actions on the work objects either remove them from her:
put aside, send back, place there, take out, take away;

then I put aside
then I have to take this signal away

or bring them nearer to her: *fetch*;

I could have fetched it

In addition to the spatial distinction between *here* and *not here*, the
worker uses another basic contrast, namely between *have* and *not
have*. Although the distinction between *here* and *not here* is closely
related to the distinction between *having* and *not having*, with
presence often implying possession as in the phrasing *now I have
the C- cards*, cards can be absent but still in possession as in *I have
the cards in the flier*.

These verbs can be arranged in a semantic field structured by three
sets of features:

location versus possession
positive versus negative versus modal
(result of) process versus state

where positive location = here, negative location = there, not here.
    For example, *put aside* differs from *get* by being concerned with
location instead of possession, from *come* by leaving the object
there instead of *here*, and from *exist* by being a process and not a
state.
    An analysis like this shows that the worker does not look at the
data from a spectator's perspective, but views them from her own
position—that of a participant—using *mine* and *here* as her points of
reference. In addition, her conception of these landmarks seems to
be positively charged, whereas *there* or *theirs* are negatively
charged, since she uses the phrasings *doesn't exist, doesn't have,
miss* about items that are located in other places or possessed by
others.
    We find this way of organizing the world quite sensible. In de-
sign it can be supported by declaring attributes of *possession* and *lo-*

cation in all system objects that designate work objects. The precise value of these attributes represents the status of the work object. To give a few examples:

> The card I am working on at the moment will be owned by me and located in a batch that is located in a job that again is located on my desk. When I send a superfluous card to the flier file,[4] its properties change: it is no longer owned by me but by the whole department, and any worker is entitled to take it if needed. Also it no longer is located in the current batch but in the flier file.

> A job can also be located in different places and owned by different people. The current job is, of course, located at my desk and owned by me, which means that no one else can see it or manipulate its items. If I cannot solve an error in the job, I can abandon it, making it a so-called *red job*[5] as in the actual system. A red job can still be possessed by me because I may want to give it another try later and therefore wish to prevent others from taking it. On the other hand, I may also want to give it up completely and transfer possession of it to my colleagues in my section. When a job is finished it becomes a so-called *green job*; I lose possession of it, and its location is changed from my desk to the area of the section. Later, after the job has been checked against the paper jobs, it is put into a box and sent to the computer department which also owns it now. It means the section can neither access nor see it any longer.

These examples illustrate how the two attributes can be used to represent important aspects of the system objects denoting work objects. In addition, the attributes have the advantage that they are easy to express visually.

A possible screen layout, shown in Figure 5, could be as follows: The location of the worker is implicitly assumed to be at the bottom of the screen, and the locations of the work objects are symbolized by their distance from the bottom and the color of the surrounding area—the higher up and the darker the background, the longer the distance.

---

4 *The flier file* is a file where workers put misplaced cards. Colleagues missing them can find them there.

5 *A red job* is an unfinished job.

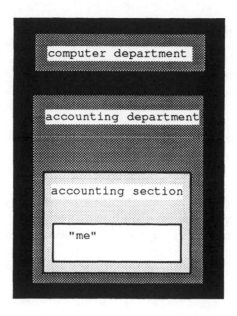

Figure 5. Screen layout based on the vocabulary of the work language.

The work objects near "me" are placed in the white rectangle in the bottom of the screen. Here is found the current job, batch, and card, together with the red jobs I want to keep for myself. At the opposite end is the computer department, where the boxes sent from the accounting department are now irretrievably outside our jurisdiction.

We have now described some properties of the system *objects* with a basis in the work language. The sentence structure of the work language can also give inspiration to the structure of the actions.

The sentences of the workers are built according to a limited set of patterns; our design can use at least the patterns relating to locating, transport, and possession. Figure 6 shows three patterns relating to transport, which occur frequently in our tape recordings.

Work Objects or Worker *move* Somewhere
Worker *moves* Work Object Somewhere
Work Objects or Worker is *placed* Somewhere

Figure 6. Sentence patterns relating to transport.

The patterns can be used for the administrative part of the work: locating, moving, and taking or losing possession of data. Here are a few examples based on the patterns shown in Figure 6:

Work Objects moving by themselves can be used if the system is to supply the next card or batch after the previous have been finished. The card or batch will "themselves" move into the work area and make themselves ready for data entry.

Worker moves Work Object will typically be used in problem-solving situations, where the worker and not the system should decide whether to move a superfluous card to the fliers or fetch a missing card from them.

To some degree, the final design will look like the direct manipulation interfaces of the Xerox and Macintosh systems. Our point is that the success of these interfaces can be explained by the fact that they have a conceptual basis in the users' work language. A natural inference is that work language is probably a rich source for ideas for other types of good interfaces.

Although we built a prototype based on the above ideas, the prototype was never tried out, and served only to make the ideas concrete. However, we feel that the design exercises at least show that it is not difficult to use work language systematically as a basis for design ideas. The benefit is that design is built on concepts which workers know, understand, and use in practice.

Perspective as a Basis for Designing Interaction Styles

Involvement Is Not Enough

The preceding design proposal is in the direct manipulation interaction style. When it is a question of performing routine tasks on objects, this is preferable, as it goes with a participant's perspective. In this section, however, we will show that this interaction style is not the *only* one. By studying the language games that arise during the process of work, one can get to the crucial problem of what user needs are, and avoid getting stuck with the (pseudo) problem of what is the superior overall interaction style (there may be none!).

The direct manipulation style is becoming more popular and is advocated by several authors (see Norman & Draper, 1986). Its purpose is to make the interface transparent, so that the users feel their work objects, and not the computer, to be present. The effect is achieved through a coding based on similarity: actions are coded as dependencies between graphical changes and user movements (e.g., clicking the mouse causes an object to invert, or dragging it causes an object to change location), whereas objects are coded as stable icons. Processes are denoted by processes on the screen, stable objects by stable pictures. The hypothesis is that similarities of this kind enhance a feeling of participation and involvement.

At least a part of the work language is suited for being coded with direct manipulation techniques, because—as emphasized in the first section—it is used for acting *in* the work situation, not for describing it from a vantage point outside. It is concerned with *concrete* instances, not with examples or classes. The theme of its sentences is *persons or work objects*, not tasks.

But our data also show that workers are not always involved participants in the work process. Sometimes they stop working, stand back to reflect, and actually adopt a spectator's perspective. On our tapes we identified conversations in which they tried to predict which kind of work processes the system would allow *(forecasting)*, and conversations whose purpose was to explain uninterpretable system behavior *(mystery solving)*. We will argue that a verbal interaction style can be preferable in these situations.

An example of forecasting is a short conversation where two workers try to fix the correct order of the tasks involved in sending a box to the computer department, entering new data into the system, closing the transmission line to the computer department, and processing cards. Mystery solving is illustrated by a long conversation where the workers try to figure out why fliers that have been removed from the flier file one day pop in again the next day.

Our thesis is that workers look at their work from two perspectives and these perspectives should be supported by different styles of interaction.

The Participant Perspective

In the participant perspective the speakers' focus of attention is on the current goal and the relevant objects; the tense of the utterances is present; their agents are the speakers; pronouns are used instead of nouns; deixis *(here, there)* is used instead of prepositional phrases *(on a table)*; the context of the utterances is known by the speakers and therefore not verbalized; the utterances often consist of simple sentences or parts of sentences; and they cohere with the work process, not with preceding or subsequent utterances.

The typical sentence in these conversations is a simple sentence (S) with a verb (V) denoting a task, and noun phrases (NP) designating worker and work objects, for example: *I must take it away*.

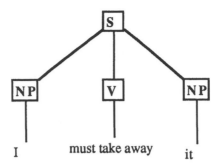

Figure 7. Participant sentence structure.

This kind of perspective can be supported with direct manipulation interaction that is intended to make workers experience the work process as a presence. We recode the workers' sentence structure in the following way: the verbs (denoting actions) are recoded as dependencies between user actions and graphical changes on the screen, while the noun phrases (denoting work objects or workers) are recoded as stable icons:

Noun phrases -> stable icons (contents: work objects or workers)

Verbs -> dependencies between graphical changes in the icons and user actions (contents: events or tasks)

Thus: User actions are coded as computer actions, and objects as computer objects.

An additional argument for having some similarity between expression and content in this case is that the computer system itself is built as a mirror image of the paper system, and the users need to determine whether the states of the two systems correspond. For example, if the system asserts that a certain job is unfinished (a "red job"), then the corresponding paper cards should be placed in a certain tray with a red header, and if a data box is sent to the computer department, then the corresponding plastic box should also be on its way. Since the system is built as a mirror image, it is important that users can easily determine whether the image is true. We believe that it will be helpful to them to design the screen as a stylized picture of the physical workplace and its objects.

The Spectator Perspective

In addition to the participant perspective, we find examples of spectator perspective. The clearest examples are found in mystery solving and forecasting. The speakers' focus of attention is no longer on the current goals of the work but on actions and relations between actions, either past events, as in mystery solving:

> But isn't it something we *cancelled, cancelled out*, we had them *from the 6th*, before, that *stayed there for two days*

or possible future events in forecasting:

> and then she said that you could reset that one, but *if* you reset it, *then* it says that the datacom is closed. Yup, and *if* it closes, *then* we can't send off our box.

The tense of the utterances is *past* (mystery solving) or *future/conditional* (forecasting); their agents may be people other than the speakers or may be indefinite; *descriptive phrases* are used instead of pronouns and deixis; the context of the utterances is presented and partially described by the speakers; the utterances can consist of *complex sentences* with subordinate clauses; and they cohere internally, not with the suspended work process.

An example of spectator perspective is found in the coordinator's speech from the section Describers and Performers. He uses complex sentences in which the noun phrases contain nominalizations with a verb kernel. Example: *Merge the data input with completion.*

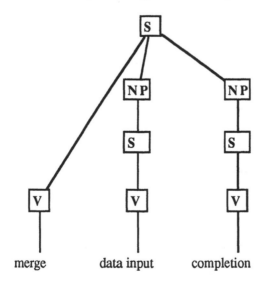

Figure 8. Spectators' sentence structure.

These sentences are very difficult or impossible to paraphrase in direct manipulation style. In the latter example, it would mean that the verb of the main sentence ("merge") should be coded by graphical changes of the coding of its noun phrases, but they are themselves verbs and should, too, be coded by other graphical changes. So the verb *merge* would have to be coded by changes of changes!

To us at least, this concept is difficult to envisage—direct manipulation seems to be like first order predicate calculus in that although it allows one to describe properties and relations between objects, it does not allow properties of properties and relations between relations. To support spectator conversation we believe that some kind of textual interaction is the most appropriate, since verbal language has a rich arsenal of devices for expressing relations of relations: subordinate sentences, nominalization, prepositions, and conjunctions.

On the basis of these data and reflections, we propose two major modifications in a redesign.

Support for Mystery Solving: Logging the Past

First, we propose that the topic of the system should not only include the present state of the thousands of cards as it does now, but also past and future actions. The former could be achieved by letting the system record a selection of the actions performed and by providing facilities for playing this log. Such a log would be helpful in the mystery solving situation mentioned earlier, where the workers cannot understand why fliers that have been removed from the flier file continue to stay in it. By playing the log backward and forward, the workers would have opportunities for discovering patterns in the fliers' behavior. At present, mystery solving conversations consist mostly of guesswork, and much time is spent on them because the workers want to understand the rules, and not just comply with them. Given a log, the conversation about fliers could run as follows. It may start just as in the tapes:

Lisa: Listen Anki, what is this?

Lisa: Two—what is this, it is a flier, I saw it was a flier.

Anki: But it was empty yesterday.

Lisa: Strange, look, 8 fliers.

Lisa: We had them before, but they disappeared.

Eva: We checked yesterday, then we didn't have any fliers left.

Lisa: I do not understand why they jump in.

But given the new facility, at this point the workers can play the log in order to verify their hypothesis about what happened since yesterday, which they could not do in the real conversation:

Lisa: Let us play the log from yesterday morning.

Eva: OK. Look there, now the fliers begin to come.

Lisa: And now they begin to disappear—look, this was mine, remember?

Eva: Four o'clock—all fliers are fetched as we thought.

Lisa: Now the date shifts—hey, look, all the old fliers jump in again!

Replaying the log has now told them that they have not made any mistakes, such as looking into the file of old fliers as the instructor in the real conversation suggested. The next step for them would be to set up hypotheses about the rules governing the behavior of the fliers:

Eva: Maybe you can only go back a certain number of days to see them.

Anki: Yes, I believe that too—maybe they just stay on for two to three days.

Eva: Let us go three days back and see what happens.

And playing the log may in fact show that the fliers jump in and stay on for three days.

   ...

Eva: It is them.

Lisa: It is those we removed.

Eva: They seem to stay on for approximately three days.

Anki: And then they disappear.

Lisa: The ones from today should have disappeared on Monday.

We recommend this method of rewriting conversations as a good technique for getting new ideas for design. It simply consists of looking at existing conversations, discovering lines of arguments that could be improved, or discovering ideas that were not explored further because the necessary information was lacking. There are two advantages in this technique:

1. It is based on real events, curbing tendencies to design facilities that have no connection with real needs;
2. It shows concretely the future uses of the facility.

The actual PGP-system does contain logs. However, one log is only for technical events in the machine, such as *Checking pool readability, poolsize = 20505600*, and is clearly addressed to technicians. Other logs are used for reporting, but although the wording to some degree is taken from the work language, it also contains technical passages, which turn out to be difficult for the users to understand and to use, as the example in the beginning of this chapter illustrates.

It is important that the facility describes the events in terms of the work process, not in terms of system features, since the data shows that workers always interpret the system's behavior in terms of work. Building the system on two perspectives amounts to creating two main subviews, both referring to the same subject, but using different concepts and means of expression.

Direct Manipulation Style

The direct manipulation style will be used in routine work such as data entry and some problem solving situations. It shows a world of work objects in which the worker can act according to the sentence patterns listed in Figure 6, two of which concern transport of work object and worker:

1. Work Objects or Worker move Somewhere
2. Worker moves Work Object Somewhere

The following figure, taken from the experimental prototype, shows the desk of one worker:

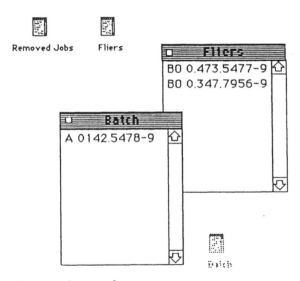

Figure 9. Screen dump of prototype.

It shows two collections of work objects, the current batch and the fliers. They are denoted by scrollable windows that can be opened and closed. Each text line of the window denotes a work object that the user can move from one collection to another by selecting it and dragging it from one window to the other.

Let us look at the action type (2) where workers move work objects. Now, although the workers distinguish between different tasks when they *talk* about work, the ideal of uniform interaction advocated in the literature seems sensible enough when they *do* work. Therefore, we want to make this kind of interaction uniform, expressing all actions in the same way, namely dragging an item from one list to another, thereby making all actions conform to the pattern *Worker moves Work Object from Source to Destination*. Thus, instead of spelling the commands by means of letters, we can use the mouse and the icons to indicate the more circumstantial *Move Document from Current Batch to Removed Documents*, for example, in the form of a menu containing *Remove Document*.

In the direct manipulation style, each "sentence" will be expressed by means of three parts:

1. Selecting: Grasping document in source list.
2. Dragging.
3. Deselecting: Letting go of document in destination list.

Recoding Direct Manipulation as Textual Style

But, as mentioned, this sentence structure will probably not support the workers when they reflect on work, so in addition to the direct manipulation style, we want the system to be able to paraphrase these actions in a textual style.

In mystery solving, a window can present descriptions of the actions done, for example, "Film number xxxx is sent to the flier file," or "The job is abandoned," or "Batch xxx is taken away."

An important argument for adding such verbal paraphrases to the graphic one is that although we may easily log the "sentences" of graphic interaction and make them available for replay during mystery solving games, the concepts they express are poorly suited for reflective conversations like mystery solving or forecasting, because they only contain a large number of simple analytical statements such as

1. Worker moves Document from Batch to Fliers.
2. Worker moves Document from Fliers to Batch.
3. Worker moves Document from Batch to Removed Documents.

In discussions, these phrasings are long and circumstantial and lack complex concepts that are good for discussing and thinking. To remedy this, we may introduce the concept *take away = move from Batch to Removed Documents*, giving (3) the shorter paraphrase

4. Worker took away Document.

A short phrase like *We took it away* is a better aid for the thought than the circumstantial *We moved it from the batch to the Removed Document file.*

However, still more complex concepts are needed. Consider, for example, the sentence

5. But they just disappeared

which a worker uses to describe the behavior of *all* fliers in the mystery solving example, condensing hundreds of statements of the form *Worker moves Document from Fliers to Batch* to one.

What is needed is not merely another kind of expression, but also a semantic structure other than that given by the direct manipulation interface. The basic linguistic problem is one of translating from one language to another, but since we want it done automatically and the system can only handle formal units, we must write explicit rules, relating the graphic expression form to one or more verbal expression forms in such a way that the referents of the two are interpreted as identical but their semantic structure as different. In short, we have to present the data from two different perspectives.

Until now, Artificial Intelligence has monopolized formal manipulation of language with the aim of making computers understand natural language. We think that this project is both scientifically and politically misguided, but on the other hand do not see any reason for not exploiting the insights of formal linguistics for other purposes. The present case is a good example of how the linguist can use all the horsepower of his or her science to achieve a different end, namely to program the machine to make paraphrases of the same data suiting different information needs of the user. Thus, the purpose is not to make the machine but the user wiser.

Solving these kinds of problems involves cases (see the beginning of the chapter) and semantic analysis, and is therefore also interesting as a linguistic problem—which we consider a prerequisite for enticing linguists to work within a paradigm other than AI or computational linguistics.

Here is a simple example of a possible rule that will translate the direct manipulation sentences from the previous section to verbal sentences:

1. the name of the list component of the selection part (part 1) is put into the source case
2. the name of the deselection part (part 3) into the destination case
3. if the middle part is dragging, *move* is put into the verb slot.

Note how cases serve as a common point of reference for both codes since they can be used to analyze both verbal and graphic signs.

This rule will translate graphic actions into sentences such as *move BO1234 from batch to fliers.*

The shorter and more synthetic phrasings used by the workers can be handled by writing rules of the type used in formal generative semantics.

For example, source = batch is the unmarked case, which is never verbalized, and the system should mirror this by deleting the source case in this situation. In addition, if the destination is the file *Removed Cards,* the work language has special words such as *take away* and *destroy* that allow the destination case also to be deleted. Thus, *Move BO1234 from batch to Removed Cards* can be shortened to *Destroy BO1234.*

Support for Forecasting: Simulating the Future

In addition to enhancing the system with a *past* tense, we suggest that adding an unreal *future* to the system would be useful in the language game of forecasting. Here the workers try to set up hypothetical dependencies between actions and exploit the modality (possible/not possible) and mode (conditional/nonconditional) systems of their language. One obvious possibility for supporting this kind of conversation is to enhance the modality of the system by adding a "conditional" to it, for example, as a small toy database, where errors or improper actions would not have fatal consequences. With such a tool, the workers could settle the correct dependencies among sending a box, entering a new date, closing the transmission line, and processing cards by experimenting, rather than by guessing as they do in the real conversation. The conversation may sound like this, given the new facility:

C: Can we close the data valve before we send our box?

D: I think it closes automatically when we have sent the box, but let us try. No—we cannot send the box when we have closed the valve.

D: But can we enter a new date before we are finished? Let us see. No, if I enter the new date, it says that the data valve closes. And when it has closed, we cannot send the last box.

The system should support the workers by shifting between direct manipulation and text style interaction. Although some kinds of situation favor one style—data entry a direct manipulation style, and mystery solving a textual style—other situations exhibit a shift from participant to spectator perspective. If an error is not too difficult to locate, the problem solving conversation can be carried out in a participant perspective, whereas other problems make the workers step back from work and reflect upon the consequences of the solution.

Conclusion

It has been important for us to show how linguistics can be used for analysis and design of computer systems in a different way than is normally done within the AI-tradition. We could sum up the approach presented in this paper by changing the DeMarco quotation a little:

> As you go about doing Linguistic Analysis, you will find yourself more and more frequently attaching yourself to the workers and following them through the work. We think of this as interviewing the workers.

Thus we not only interview workers in the coffee room, but we are more interested in the way they talk during work. The reason for this is that we believe that work language is not fortuitous, but based on collective and partially unconscious knowledge and experience.

The result is *our* interpretation of these hidden patterns, and can be used as a basis for prototyping and discussions with the users, making them aware of patterns of work practice and conceptual structures they are not paying attention to during daily work.

References

Andersen, P. Bøgh & Holmqvist, B. (1986). A toolbox for analyzing work language. In J. D. Johansen & H. Sonne in collaboration with H. Haberland (Eds.), *Pragmatics and Linguistics*. Odense, Denmark: Odense University Press.

Andersen, P. Bøgh. (1990). *A theory of computer semiotics. Semiotic approaches to construction and assessment of computer systems*. Cambridge, UK: Cambridge University Press.

Fillmore, C. J. (1968). The case for case. In E. Bach & R. T. Harms (Eds.), *Universals in linguistic theory* (pp. 1-90). London: Holt, Rinehart & Winston.

Fillmore, C. J. (1977). The case for case reopened. In P. Cole & G. M. Sadock (Eds.), *Syntax and semantics: 8. Grammatical relations* (pp. 59-81). New York: Academic Press.

Holmqvist, B. & Andersen, P. Bøgh. (1987). Work language and information technology. *Journal of Pragmatics, 11*, 327-357.

Holmqvist, B. (1989). Work language and perspective. In *Scandinavian Journal of Information Systems, No.1*. Aalborg, Denmark: University of Aalborg.

DeMarco, T. (1978). *Structured analysis and systems specification.* New York: Yourdon.

Norman, D. A. & Draper, S. W. (1986). *User centered design— New perspectives on human computer interaction.* Hillsdale, NJ: Lawrence Erlbaum Associates.

Ullman, S. (1962). *Semantics. An introduction to the science of meaning.* Oxford, UK: Basil Blackwell.

6

Workplace Cultures: Looking at Artifacts, Symbols and Practices

Keld Bødker and Jesper Strandgaard Pedersen

Often system designers encounter the consequences of culture when an implemented system is not used as intended or does not live up to the users' expectations. This is obviously too late a recognition of the importance of the culture existing in workplaces. We advocate the study of workplace cultures in relation to design of computer systems. In this chapter we give examples and guidelines for how to study workplace cultures.

In one project, three designers, in collaboration with journalists, conducted a survey to study the possibilities of computer support of editorial and administrative tasks in the editorial section at a Danish radio station. During the survey, it became clear to the designers that some of the ideas for computer support envisioned by the journalists conflicted with the conditions established by the organization. It was also difficult for the journalists to formulate explicitly the need for changes in work organization, for example, where computer support was considered necessary. Using elements of a study of workplace cultures, the designers were able to identify values and beliefs relative to work situations which showed that the envisioned computer support would be unrealistic. Furthermore, both the designers and journalists were able to use these observations to identify problems with the proposed design. This example will be developed below.[1]

Often in theory and work practice it is believed that the "irrational superstition" of users is the perceived foe, causing resistance to change. From the system designers' point of view, what seemed important to the users was not found relevant and vice versa. In this chapter we take seriously the apparent irrationality and missing logic that is said to exist in workplaces. We then advocate a study of workplace cultures in relation to the design of computer systems (Bødker, 1990).

What is Culture? Beyond the Quick Fix

We have probably all experienced the effects of workplace culture, either when we have changed jobs and entered a new organizational setting, or when we have met and worked with people from other organizations. What we often experience is the sense that "they do things differently and in awkward ways, which to us seem irrational and with no purpose." Nevertheless, after some time—it may take from a few months to several years to learn the code of conduct—it appears that we find this behavior meaningful, rational, and purposive, and no longer strange at all. We are gradually socialized and trained to understand and accept the actions, practices and meanings in the organization.

This idea focuses our attention on different applications of the culture concept. Some authors talk about organizational culture; here, to be more precise, we talk about workplace cultures. The distinction between the two, to us, is related to the size and proximity of the object of analysis. The point is that we use the concept of workplace cultures for a relatively small working environment, where it makes sense to talk about the workers as an entity. This implies, however, that in certain situations, when studying small organizations, the workplace culture and the organizational culture are identical. In other situations, the values and beliefs in the working environment are distinct from corporate culture, corporate ceremonies, espoused values, slogans, etc.

We conceptualize culture to be a system of meaning that underlies routine and behavior in everyday working life. Various departments or workgroups develop their own artifacts—like dresscode and decorations; their own language—like jargon and metaphors; and their own work practices—like routines and modes of cooperation (Borum & Pedersen, 1990). What we are looking for is aspects of

---

[1] In the examples given in this chapter names and nonessential details have been changed or left out in order to preserve the anonymity of participants.

the culture developed in working groups or by groups of employees handling similar work situations, but perhaps at different places in the organization.

The central purpose of our approach to the study of workplaces as cultures is to capture and understand the values and beliefs existing in workplace settings. It is the *shared* values and beliefs, constituting collective understandings, which are of interest to us in a cultural study. Smircich (1983) implies that analysis of cultures should involve multiple points of view and the relations between them, when she says "An analysis of an organization as a culture must go beyond any single individual's understanding of the situation" (p. 162). Studying organizations or workplaces as cultures by looking at artifacts, symbols, and work practices is one way of trying to grasp and understand the intangible phenomenon that others have identified and labeled "climate," "spirit," "milieu," and so forth.

Roots in Anthropology

The concept of culture used in organizational studies originates from anthropology and has only recently been related to an organizational context. Starting in the 1950s and continuing through the 1960s and 1970s, various authors have used the anthropological concept of culture and related methodological issues to describe and analyze organizational behavior. In the 1980s, it has become widespread to analyze and describe organizations in a cultural perspective; the literature contains a vast number of books and articles on this issue (Pettigrew, 1979; Administrative Science Quarterly, 1983; Frost et al., 1985; Schein, 1985; Meyerson & Martin, 1987; Van Maanen, 1988). But what is the relation between the anthropological concept of culture and the concept of culture applied to organizational studies? What do we really mean when we talk about the "culture" in workplace settings? We hope that the discussion in this chapter addresses these questions.

Within anthropology, "culture" has been used to denote the particular "way of life" of foreign societies. The life of bushmen in the Kalahari desert in South Africa, or of headhunters in Southeast Asia, have been described by anthropologists under the headlines of "culture," meaning a description of their world view and life forms.

This presentation points out that culture is a *relative* phenomenon. The way of life of one group of humans is described as being distinct from others, and this distinctiveness is emphasized and depicted as being specific to this group in relation to other groups. The anthropologist C. Geertz very precisely expressed what characterizes an interesting cultural analysis, stating that "understanding a people's culture exposes their normalness without reducing their particularity," (Geertz, 1973, p. 14). Closely related to this observation is a classic discussion within anthropology of the potential danger of ethnocentrism. That is, instead of describing the foreign tribe or society, the anthropologist describes his or her own culture by describing only that part of the stranger's way of life that contrasts with the anthropologist's way of life, further explaining and making sense by using roles and codes from his or her culture.

Culture in Organizations and Workplaces

When we apply the concept of culture to understand workplace behavior we look at a workplace setting as if it was a foreign tribe or society. We see it as a community with some integrity, some sense of its own identity, and some common artifacts, symbols, work practices, and underlying values and beliefs. We try to capture the uniqueness of this part of the organization in our attempt to understand the meanings attributed to the established behavioral patterns, workroutines, and symbols and artifacts used in the workplace.

The workplace is seen as *being* a culture. The values and beliefs of the culture are understood to have grown out of experience, and are conceptualized as a system of meanings underlying artifacts, symbols, and work practices. Although referred to as a system, the culture is not explicit but implicit; that is, hidden behind or in the various artifacts, symbols, workroutines, and established patterns of cooperation. The culture is only made explicit when expressed in words or actions; for example, explication takes place when an outsider or newcomer, by asking questions, forces the workplace members—the cultural insiders—to reflect upon the reasons behind a particular artifact, symbol, or work practice.

The introduction of a cultural perspective on organizations and workplaces has meant a change in orientation and focus. In the past, organizational scientists, when trying to describe and understand how organizations operated, were preoccupied with organizational structure, division of labor, hierarchies, competence and lines of communication (see Burns & Stalker, 1961; Leavitt, 1965; Minzberg, 1983). When applying a cultural perspective on organizations and workplaces, researchers and practitioners now emphasize interpretive capacities and such features of working life as language, traditions, rituals, myths, and stories. A cultural perspective directs attention to the symbolic significance of what is normally regarded as the most rational (i.e., "natural") aspects of working life (Morgan, 1986). A cultural perspective also shows that an organization is not just structures, diagrams, lines of command, and division of work, but also shared systems of meaning, (Pedersen & Sørensen, 1989).

A cultural approach to the study of organizations and workplaces looks at well-known organizational phenomena in new ways. For example, a meeting is seen as much more than information distribution, decision making, and reporting. The meeting is seen as a *ritual* where status, power, and importance are distributed, and sense is made out of confusing events. A meeting viewed as a ritual, therefore, has an *instrumental* as well as an *expressive* side to it.

Artifacts and symbols are seen as objects which synthesize and express meaning. Thus, a symbol represents something more than itself. The complex and somewhat chaotic nature of "reality" is mapped, condensed, and expressed in the artifacts and symbolic systems. *Meanings* do not, however, exist in artifacts, symbols, or practices. They are *assigned* to artifacts, symbols, and practices by people who perceive and interpret their content and context (Smircich, 1983). In this way, repeating an almost classic statement, we say that the organizational reality is socially constructed (Berger & Luckmann, 1966). We have chosen artifacts, symbols, and practices as our entrance to culture, because they are tangible, accessible, and visible. Nevertheless, identifying the meanings associated with these artifacts requires interpretive skills. Cultural manifestations are easy to obtain but difficult to interpret, because they are ambiguous and may hold multiple meanings and understandings.

Many anthropological concepts have been presented in various studies of organizational and workplace cultures. They are so numerous and varied that we are not able to give a full account. The concepts that we have found to be the most useful for studying and analyzing workplace cultures from the perspective of systems design are listed below. These concepts have their roots in anthropology and organizational sociology.

• *Physical and material artifacts*: office layout, decoration, worktools, dress code

- *Verbal symbols*: stories, sayings, jargon, anecdotes, metaphors
- *Work practices*: work routines, modes of co-operation, gestures, rituals

Our entrance to a study of workplace cultures is through artifacts, symbols, and practices. The idea underlying our approach to a study of workplace cultures is that the core values and main beliefs constituting the culture are *manifested* in various material and physical artifacts, like office layout, decoration, work tools, and dresscode; *expressed* in verbal symbols, like stories, sayings, jargon, metaphors, anecdotes; and *dramatized* in work practices, like work routines, modes of cooperation, gestures, rituals. By using these manifestations of culture as "handles," we can approach values and beliefs by means of a collaborative dialogue into the meaning ascribed to artifacts, symbols, and everyday work practices.

These artifacts, symbols, and practices thus serve as a kind of 'collective storage' for workplace knowledge and experience in relation to the workplace members (Levitt & March, 1988). This workplace memory, which carries the historically developed workplace experience, is activated under certain circumstances and passed on to members and to new generations of members as a "fabric of meaning," producing and reproducing collective understandings.

Often the most accessible symbol system is language. By attending to what people say, we can learn how they organize their perceptions of their working life and make sense of their experiences (Smircich, 1983). Approaches focusing on language use in workplace settings have been discussed by Eleanor Wynn in Chapter 3 and by Berit Holmqvist and Peter Bøgh Andersen in the preceding chapter. In this chapter we will focus on such cultural forms as artifacts, other verbal symbols, and work practices.

Paying Attention to the Process

Concerning the approach to studying workplace cultures, two extreme views exist, as indicated by Beyer and Trice (1987). One extreme holds the view that "an organization's culture is so obvious it can be immediately sensed by outsiders when they step in the door." The other extreme argues that, "an organization's culture is so elusive it can be revealed—and then only partially—only by outside experts after a lengthy study" (p. 5). Our approach is an attempt to maneuver between these two poles.

The problem we face as designers—and cultural outsiders—is that we cannot just enter a workplace setting, sit down, observe the culture, go home, write up the results based on these observations, and then come back and present the results: "Here you are, this is how your culture looks"! This is not possible because we come to the workplace setting as cultural outsiders and because meaning cannot be observed. We can of course enter a setting and observe the life as it appears—and we should do so—but to the extent that we meet the cultural system as outsiders, we do not understand the culture and behavior of the cultural insiders; the meaning is still hidden.

We do *not* think it is a simple task to engage in a study of workplace cultures in a given setting. When we have entered workplaces, people seemed to agree on the existence of a culture; they might talk about "the culture here" and appear to recognize the importance of being aware of the particular composition and expression it has. However, when asked further, only few have a precise image of what *it* is and where to look for it. This is not to say that these people are ignorant, but to point out that tracing cultural elements is difficult and requires extensive effort, as culture is a notoriously complex and ambiguous concept.

For these reasons a collaborative approach is important to us, wherein the analysis takes place in an interplay between cultural insiders and cultural outsiders. We perceive a cultural study to require a *collaborative* approach, as opposed to an *expert* approach. In a collaborative approach, interpretation of artifacts and symbols, elicitation of the shared meaning underlying the daily routines, as well as decisions about how to use the results, are carried out in close cooperation between committed outsiders and open-minded insiders.

A Collaborative Approach

We propose the process depicted in Figure 1 for studying workplace cultures. Below we briefly explain the various steps of a cultural study with a collaborative approach as they are identified in the figure. By its very nature, such a description is analytical.

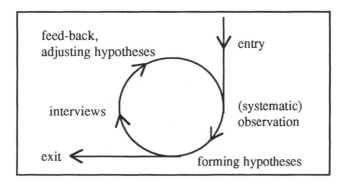

Figure 1. The process of conducting a cultural study.

To the outsider *entering* a workplace or an organization, the daily life appears chaotic, decisions made by organizational members seem uncoordinated and unmotivated, and many activities appear to be confusing and irrational. As soon as cultural insiders start explaining the "logic" underlying these actions, though, one understands much more of what is going on. However, entry is very important for the cultural outsider and what seems strange and puzzling should be recorded at this initial state.

From *observing* the routines and procedures of work in a part of an organization, we can *form tentative hypotheses.* Typically we notice a particular jargon or mode of cooperation that everyone takes for granted. But we can only get certain types of information from observing the routines and procedures of work in a workplace setting. Consequently, we get supplementary information by collecting other types of data, and we *interview* people, in more or less formalized interviews utilizing more or less structured interview guides, typically based on the ideas generated by previous observations. By presenting the observations and our tentative hypotheses to the insiders at the interviews we get *feedback,* and we can *adjust the hypotheses.* This process is continued until *exit* is made, providing for a final analysis and, eventually, a description of the workplace culture.

Culture is a collective phenomenon and, as explained earlier in this chapter, of interest in a cultural analysis are the values and beliefs that are shared among the workplace members. We are only interested in studying those values and beliefs which have implications for interpersonal relations and activities at that particular workplace. This indicates that both individual interviews and group interviews are appropriate in a cultural study. A group interview or discussion organized around, for example, videotaped work situations or other forms of documentary accounts, are situations where outsiders can get information about and study the shared definitions of reality. Another way is to conduct individual interviews with organizational members, and afterwards compare the results across the individual interviews, examining the ways they agree, and looking for common stories.

It is crucial to us as cultural outsiders to work with insiders in order to grasp the deeper meaning of the symbols and activities taking place in that particular working environment. We can recognize this and aim to see the world as the workplace members see it, to learn the meaning of symbols, actions, and events, which the members draw from those forms and activities. We can approach the shared meaning through "situated" or "occasional inquiries," as described by Lucy Suchman and Randy Trigg in Chapter 4; that is, while being present as the work goes on, we ask questions related to specific events. In general, we find that the cultural forms are very helpful entrées to the culture, as they provide an opportunity and occasion for dialogue. To do this, we have to engage in close interaction based on informal dialogues with cultural insiders.

Using Cultural Forms as Handles

The outsider or newcomer must have an explanation as to why a particular material artifact is placed on a particular spot. What is the story behind it? What does it mean to the members in the workplace? For example, during one study, we noticed a special decoration standing on a pedestal in the lobby of the company. Looking at it more closely we saw that this metal-like item was some kind of technical part belonging to one of the company's machines. It was put on the pedestal and behind glass, like a precious jewel or a rare painting. Below the glass, a small metal plate read "Thompson's first pump valve for RF-Turbines."

The meaning of this artifact was unclear to us, so we asked what it meant and why it was on exhibit. We were told that their company for many years had been solely dependent on one supplier of pump valves for their production of turbines. Even though this supplier was reliable and trustworthy, they were uneasy about being so dependent on one supplier. Eventually another company started producing a pump valve which RF-Turbines could use for their production, thereby becoming less dependent on their former supplier. As the metal plate indicated, the artifact exhibited in the lobby was the first pump valve from their new supplier. To the members of the workplace it symbolized their "liberation from captivity" and the beginning of an era of economic prosperity. This artifact and the meaning assigned to it proved to be an insightful entrance to that particular workplace. By explaining the story and meaning behind this artifact, the workplace members passed on important information to us about core values and prevailing beliefs in their culture. For example, it became clear that members of the culture valued autonomy and independence, and had a strong belief in self-determination.

As we have mentioned earlier, the trap to avoid in holding the position of cultural voyeurs is that of ethnocentrism. By this we mean that we as outsiders project our ideas onto the subject of the study (the cultural insiders), ascribing certain motives and imposing our ideas of rationality on their behavior. We have personal interpretations of what is happening, and engage in a process of postrationalization, without examining the interpretations the cultural insiders attach to these phenomena. An example of this would be if a person noticing the RF-Turbines artifact had not bothered to ask the workplace members what its meaning was, but instead had chosen to interpret this exhibition as a sign of the company being proud of their own product.

Symbols provide an opportunity to communicate a message and express norms about how to behave or not behave in certain situations. They articulate beliefs about what is considered important or unimportant in relation to the daily activities in the working environment. In another study, we found that the culture was characterized by a strong work ethic, oriented toward hard work and results. The distinction between leisure time and working hours was blurred; working late was expected and associated with being serious and committed; status among workplace members was highly and positively correlated to how much time they would spend on their work.

Our key to this culture was through a verbal symbol, a saying, which we repeatedly encountered. The saying was phrased: "to walk around with your parking lights turned on." We were a bit puzzled by this saying and considered various interpretations. Did it, for example, mean that you were bright or dim? It soon became clear to us that the saying was not associated with anything positive; on the contrary, it was considered improper to "walk around with your parking lights turned on," as the phrase was used as a metaphor whenever somebody became abstract, intellectual, or high-flown.

When we explored further, we learned that the saying originated in a special situation, which had become a widely told story. The story was about a former employee who was recruited directly from The Technical University. The new employee did not live up to expectations: "He went around all the time with his parking lights turned on," indicating that he was utilizing his batteries for the wrong purposes. He was considered too "academic," speculative, and not productive, thereby failing to meet the hard work ethic. From this story, another aspect of this workplace culture became clear to us. We discovered that formal education and degrees were not considered important, as the culture was dominated by people who were autodidacts, those who thrived on learning-by-doing and who valued visible, concrete results, instead of formal education.

By these examples we hope to have illustrated how artifacts, symbols, or practices provide excellent opportunities to get rich information on shared values and beliefs in a workplace. However, the problem is that artifacts, symbols, and practices are not easily deciphered or decoded; they are open to many interpretations. Furthermore, artifacts and symbols are relative, implying that their message is invoked and defined by the specific context. The point is that norms and practices are not universal but context-dependent.

As outsiders, we need to study our interpretations and images of the symbols with cultural insiders in order to understand their underlying meanings.

Workers as Cultural Insiders—Implications for Design

To illustrate some potential implications of a study of workplace cultures for the design of computer systems, this section further elaborates the example presented in the introduction. The example is taken from a project in an editorial section at a Danish radio station. The editorial section produces a daily program with political and cultural comments on national and international issues.

As the first part of the project, a survey on the applicability of computer support of editorial and administrative tasks was carried out. The survey was done by a group working part time and consisting of three journalists, a secretary, and three designers. The group was established with the objective of studying the possibilities for computer support of the planning and administrative follow-up of the program production. They were to report within three to four months in a 20-page memo and with a prototype demonstrating screen images, etc.

We, the designers, started the survey by following the daily production of programs. We observed editorial meetings, followed journalists doing "research and interviews in the field," observed journalists work in their offices and in the studios producing taped features, and we observed the two journalists planning, producing and reporting the live program. This area was completely new to us and we spent several weeks getting acquainted with the workplace.

We documented our observations with graphical illustrations on large sheets of paper, forming a 1 x 2 m wall chart, depicting the work situations with the journalists in different roles, using various information and tools. The wall chart was presented to the working group to provide feedback, to check "if we had understood it right," and as a reference to discuss inexpediencies which we had observed.

Through this discussion, we identified a number of areas where computer support was considered applicable. These areas were then further studied and discussed to provide input for the design of a simple horizontal prototype. The horizontal prototype, consisting of empty screen images for all the areas of the design proposal, as well as a vertical prototype detailing the functionality within one area—that of the retrieval of information on notes and sources—were then developed. The prototypes were demonstrated in the working group and evaluated by relating the intended use-situation to the wallchart descriptions of the work situations. The vertical prototype was further tested and evaluated by a journalist from the working group, who used it in his daily work.

At this point we felt doubtful about parts of the design. During our study of the work situations we had made observations that we thought might invalidate parts envisioned by the working group, namely parts that were based on idealistic perceptions of work situations.

In order to explore these uncertainties further, we returned to studies of work situations. One approach taken was a study of workplace cultures focusing on values and beliefs in relation to central, individual aspects of the journalist's work: the journalist doing general research in a new area or research for a particular feature; the journalist's use of notes and sources; and the journalists' views on what constitutes a good feature and a good program. Thus, the artifacts that were used as handles in this study were *work tools* such as the journalist's note pads; *work routines* such as his or her use of sources; and the *products of work* such as the individual feature, and the program as a whole.

The results of these studies were presented orally and were used to stimulate and provoke a discussion of the premises and consequences of the design. The results were condensed into 11 assertions of how we, as designers and outsiders, perceived the work situations *to be,* as opposed to how the journalists and the secretary, as insiders, said or thought they *were.* In a number of areas, the results of this study in terms of cultural values and beliefs clearly conflicted with the values formulated implicitly or explicitly as part of the envisioned work situations.

For example, when discussing design ideas, the journalists highly valued computer support for retrieval of notes and sources. Given their use of note pads, the study made it clear that only a very small part of their notes might end up in the computer system. The journalists take notes at many different places, during interviews, at meetings or seminars, or at home, and they don't have time to rewrite them. So we could formulate an assertion saying "notes remain handwritten." From this, it was clear that there would only be a very limited amount of information to retrieve, unless the journalists individually chose to change their way of taking and keeping notes.

As to the journalists' view on sources, the study showed that they consider their sources to be an integrated part of themselves and thus highly private. This meant that for reasons of privacy, facts about sources should never become part of a computer system, not even in a private file. Again we could formulate an assertion, this one saying "sources are highly private," implying that what could become part of the computer system were the names of more formal contact persons, which the secretary already kept in a manual file.

Following a presentation of the final design, we presented the assertions, formulated as short, provocative statements, in order to *challenge the design.* The objective was to test if—and where—the design was founded on *espoused or idealistic assumptions* about the work, which were not reflected in the actual work practice. During the discussion we constantly focused on contradictions or tensions between how the journalists and the secretary said or believed they worked, and our interpretation of how they worked. We could then discuss changes of the design or become aware of critical premises and fundamental values and beliefs regarding changes of the work practices. In turn, these findings would have to be regarded as essential organizational decision points if the new system was to be used as intended.

In summary, the detailed study of work situations and the comparison of the results with the design proposal identified a number of prerequisites to the proposal in terms of decisions at organizational and individual levels, as well as changes to the design proposal.

The Insider–Outsider Dilemma

Conducting a cultural study as a designer and a cultural outsider requires a delicate balance; several conflicting considerations have to be handled with utmost care. Getting to know about a culture is a two-edged sword. As we dig deeper into the culture by getting more information and explanations about the patterns and symbols expressing it, we are at the same time in the midst of a socialization process ourselves. We are learning about the cultural traditions, norms, values, and beliefs existing at that workplace. During this process of socialization, we learn to take certain explanations for granted. Behavior that appeared mysterious to us when we first entered this setting later seems normal and legitimate. We accept certain features of life and gradually become less aware of their existence, responding more or less unconsciously to them.

The trap we are likely to be caught in is one where we suddenly find ourselves in a position unable to recognize, distinguish, question, and thus, reflect on culture: It is just there. An immediate risk exists that we get so "culturized" that we "cannot see the forest for the trees."

Consequently, the designer doing a cultural study faces the dilemma on the one hand of getting enough knowledge to understand the meaning of the cultural forms (and thereby the cultural pattern as such) and, on the other hand, of keeping the necessary perspective on the object of study.

In this discussion we would also like to draw attention to an awareness of the role of the cultural outsider as someone who may influence the system of meaning being studied. When we contest the patterns of tradition and routine activity, they tend to become more active and observable. Some claim that culture is only revealed when change occurs; one may then see a specific feature of the culture. As pointed out by Garfinkel (1965), culture is revealed through deviance and anomalies; when people act in ways the system does not expect. As outsiders we should inquire about the behavior we observe in a working environment. We will be given an explanation, and the logic underlying this behavior then brings us closer to the beliefs and key values guiding members of this culture. To attain a deeper understanding of the dynamics of behavior, the next step can be to challenge the explanations provided by the cultural insiders. For example, by provoking without being insensitive or abusive, we can challenge the explanation by indicating other explanations and observations of behavior that point to different interpretations, in order *not* to accept the first explanation given.

The cultural study generates information that may challenge the image held by organization members; it is not a neutral process. Nevertheless, this is not unique for a cultural study, but something which is true for all types of approaches, whether in the form of a report of research findings with tables and figures, or as an interactive feedback session conducted by consultants or researchers. In fact, all researchers engage in interpretive acts in the translation of data into meaningful forms. All data, whether verbal, statistical, or pictorial, are meaningless when abstracted from the context. Data only become significant when they are made sensible and coherent through the mediation of human meaning (Smircich, 1983).

Concluding Remarks

The collaborative approach to the study of workplace cultures presented in this chapter is important, because through this process the cultural insiders become aware of the patterns, and designers may come up with a better system design based on an understanding of the central values and assumptions at the particular workplace. By becoming aware of the meanings ascribed to actions and symbols, we—as actors in workplace settings—recognize our active role in the construction of the shared reality and gain a chance to empower ourselves in our workplaces.

Acknowledgments

We thank Susanne Bødker, Lucy Suchman and the Editors for their comments on earlier drafts.

References

Administrative Science Quarterly (1983, September). *28* (3).

Berger, P. L. & Luckmann, T. (1966). *The social construction of reality*. New York: Doubleday.

Beyer, J. M. & Trice, H. M. (1987, Spring). How an organization's rites reveal its culture. *Organizational Dynamics, 15* (4), 5-24.

Borum, F. & Strandgaard Pedersen, J. (1990). Understanding IT people, their subcultures, and the implications for management of technology. In F. Borum & P. Hull Kristensen (Eds.), *Technological innovation and organizational change–Danish patterns of knowledge, networks and culture* (pp. 219-248). Copenhagen: New Social Science Monographs.

Burns, T. & Stalker, G. M. (1961). *The management of innovation*. London: Tavistock.

Bødker, K. (1990). A cultural perspective on organizations applied to analysis and design of information systems. In G. Bjerknes, B. Dahlbom, L. Mathiassen, L. Nurminen, J. Stage, K. Thoresen, P. Vendelbo, & I. Aaen (Eds.), *Organizational competence in system development*. Lund, Sweden: Studentlitteratur.

Frost, P., Moore, L. F., Louis, M. R., Lundberg, C. C., & Martin, J. (Eds.). (1985). *Organizational culture*. Beverly Hills: Sage.

Garfinkel, H. (1965). *Studies in ethnomethodology*. Englewood Cliffs, NJ: Prentice-Hall.

Geertz, C. (1973). *Interpretation of cultures*. New York: Basic Books.

Leavitt, H. J. (1965). Applied organizational change in industry: structural, technological and humanistic approaches. In J. G. March (Ed.), *Handbook of organizations*. Chicago: Rand McNally.

Levitt, B. & March, J. G. (1988). Organizational learning. *Annual Review of Sociology, 14*.

Louis, M. R. (1985). An investigator's guide to workplace culture. In Frost et al. (Eds.). *Organizational Culture*, op. cit.

Meyerson, D. & Martin, J. (1987). Cultural change: An integration of three different views. *Journal of Management Studies, 24/6,* 623-647.

Minzberg, H. (1983). *Structure in fives: Designing effective organizations.* Englewood Cliffs, NJ: Prentice-Hall.

Morgan, G. (1986). *Images of organization.* Beverly Hills: Sage.

Pedersen, J. Strandgaard & Sørensen, J. S. (1989). *Organizational culture in theory and practice.* Aldershot, UK: Avebury/Gower.

Pettigrew, A. M. (1979). On studying organizational cultures. *Administrative Science Quarterly, 24,* 570-581.

Schein, E. H. (1985). *Organizational culture and leadership–A dynamic view.* San Francisco: Jossey-Bass.

Smircich, L. (1983). Studying organizations as cultures. In G. Morgan (Ed.), *Beyond method. Strategies for social research* (pp. 160-172). Beverly Hills: Sage.

Van Maanen, J. (1988). *Tales of the field–On writing ethnography.* Chicago: University of Chicago Press.

Van Maanen, J. & Barley, S. (1985). Cultural organization: Fragments of a theory. In P. Frost, L. F. Moore, M. R. Louis, C. C. Lundberg, & J. Martin (Eds.), *Organizational culture* (pp. 31-58). Beverly Hills: Sage.

Part II

Designing for Work Practice

7

Setting the Stage for Design as Action

Susanne Bødker, Joan Greenbaum, and Morten Kyng

Around the sixteenth century, there emerged in most of the European languages the term "design" or its equivalent. The emergence of the word coincided with the need to describe the occupation of designing. ... Above all, the term indicated that designing was to be separated from doing.

Cooley, 1988, p. 197

Equipped with the discussions in Part I on the importance of the practice of the users and how to get a better understanding of it, we now turn our attention to the practice of design. Our main concern is how to set up design processes in such a way that users may participate directly in those processes; in other words, how to design cooperatively with users.

The First Scene: Sticking to Explicit Knowledge

In this scene we encounter a large Danish software company that wants to build a system to control waste incinerator plants. The software department picked a group of their best designers and asked them to do the job. These designers were mostly programmers and engineers who were quite experienced in technical analysis and programming. They started out analyzing how switches, valves, control panels, meters, and displays worked in the current manually controlled system for incinerator plants. They talked with the engineers who had built the current system, in order to collect information on how the physical incinerator system worked. They wrote descriptions and drew diagrams showing how measurements were made and how various conditions such as extreme temperature or excessive smoke were indicated and controlled with the switches on the control panel. The purpose of all these analyses was to identify a complete set of rules describing how incinerators were controlled and thereby how the new expert system should work.

Although the designers were quite competent engineers and programmers, they found that it was difficult to design the rules for the expert system. One designer asked, "Why don't we ask the operators who are experienced in controlling the incineration how they do it?" The group agreed this was a good idea and a visit to an incinerator plant control room was arranged.

During this visit one of the designers asked an operator the fifty dollar question: "How do you really use these displays and switches to control the incineration?" The operator hesitated for a moment and answered: "Ehh... I really don't use the displays that much, they mostly serve as a kind of background information... but I use this TV monitor pretty much to see the size and the color of the flame." He continued: "For instance, an hour ago, the flames got too big—then I turned this handle [indicating with his left hand] a little counter-clockwise, and dragged this other handle a little in this direction [indicating with his right hand]. The flame size was not reduced fast enough and I then closed the air valve completely for a few seconds by pushing this button [pointing]—but that is rarely needed." The designers were quite surprised by this explanation—they had never considered the TV-monitor important for the control activity.

The story illustrates that if the kind of insight the operators have is not brought to bear on the design, the result will be systems that don't fit the situations for which they are intended; systems which may even run contrary to competent performance. In this case the designers realized the importance of some kind of user involvement in the design process, but they still only considered users as sources of *explicit knowledge* which could be communicated and then formalized into rules.

In interpreting the story, however, *we* see strong arguments for a more direct and active involvement of users in the design of computer systems. Essential parts of the users' knowledge are embodied in their *involved and unreflected performance*. This knowledge is not confined to the users as performers, but also exists in their relation to the environment, with its monitors, handles, dials, and other tools. Thus, the changing picture of the flame on the monitor *triggers* a reaction such as the turning of a handle, but the operator is not conscious of exact details of the picture. He probably could not make them explicit even if he wanted to. Verbal accounts such as the one given by the operator are obviously only a shallow representation of his involved performance in his work.

The Second Scene: Description versus Experience

The obvious "solution" to the problem of how to bring the knowledge and skill of the users to bear on the design is to involve the users directly in design work itself. This, however, is not as simple as it may sound, as illustrated by the following scene from the UTOPIA project.[1] The project was formed cooperatively by the Nordic Graphic Workers Union and research institutions in Denmark and Sweden. The aim was to design computer-based tools for text and image processing. To this end a design group consisting of skilled typographers and designers with a background in computer science was set up. In the first activities of the project, both end users and designers played active roles in the mutual learning process: teaching, discussing, and learning about their own work and that of the others in the group. However, when they moved to design activities in terms of writing "traditional" system specifications, the designers took the initiative as we shall see in the following example.

> Late one afternoon, when the designers were almost through with a long presentation of a proposal for the user interface of an integrated text and image processing system, one of the typographers commented on the lack of information about typographical code-structure. He didn't think that it was a big error (he was a polite person), but he just wanted to point out that the computer scientists who had prepared the proposal had forgotten to specify how the codes were to be presented on the screen. Would it read "<bf/," or perhaps just "\b" when the text that followed was to be printed in boldface?

It was back in the early 1980s, when almost all commercial text-processing systems were based on text screens and used codes like the ones mentioned in the example. Nevertheless, the interface which we had been discussing for hours was of the "WYSIWYG" (What You See Is What You Get) type using a graphic screen. No codes were to be shown, and letters to be printed in bold would simply be shown in bold on the screen. What was going on? Had the typographer missed the point?

---

[1] The Utopia project is one of several Scandinavian projects where shop stewards and other workers cooperated with researchers and designers on evaluation and design of computer systems. An overview of the projects and related activities is given in the paper "The collective resource approach to systems design" by Pelle Ehn and Morten Kyng. The Utopia project is described by Bødker, Ehn, Kammersgaard, Kyng, & Sundblad in the paper "A utopian experience—On design of powerful computer-based tools for skilled graphic workers." Both papers appear in Bjerknes, Ehn, & Kyng (1987).

The typographer who made the comment had certainly been paying attention during the presentation—that's how he noticed that the codes were not mentioned. He understood computers in newspaper production very well; in fact, he had been responsible for the education of several hundred typographers during the installation of a new computer system at the newspaper where he worked. However, the written and oral presentations of the designers were not able to bridge the gap between his experiences and the new design. The design situations were too unfamiliar to him, because they contained no *clues* as to how to apply his experience and knowledge as a typographer. The designers tried a number of other approaches, including scenarios describing possible future work organizations and computer-based systems, but with limited success. Then they began to use mock-ups or simple cardboard representations of what the system might look like (see Chapter 9). This changed the roles dramatically: Cooperation between designers and skilled workers shifted from situations where the workers usually agreed on descriptions made by the designers, into working sessions where (simulated) computer-based tools were tried out in practice and actively critiqued. The shift was from situations unfamiliar to typographers to situations that had some *family resemblance* with their work; situations where the mock-ups functioned as reminders, triggering work operations based on the *tacit, non-explicit knowledge* of the workers. This was a step toward *design-by-doing*. The mock-up simulations allowed the users to draw directly on their skills in carrying out work in the application area. In this way *doing* became a primary activity in cooperative design.

However, when using mock-ups, unreflected performance—those activities we don't think about—doesn't last long. Frequently *breakdowns* occur in the work, usually because certain ways of doing parts of the work process were not possible when using the mock-up. In such situations the mock-up changes from being "transparent," almost like our use of our hands, to being present as the focus of attention. Such a breakdown thus changes the design situation from one of using the mock-up to one of discussing computer support for its use.

Thus, a breakdown brings about a discussion resembling system description discussions, but at a much deeper level. The difference is that after a breakdown, the group can specifically discuss the work that led to the breakdown and identify possible causes based on the *use* of the mock-up. For example, a user may suddenly stop her work on a mock-up because she felt that she had to move her hands too often between the mouse and the keyboard. Based on this knowledge about the use of the mock-up (the sequences of operations, etc.) the design team may continue analyzing the set of operations, revising it to better fit the use situations. This knowledge from use is exactly one of the keys that is often lacking in traditional system descriptions.

Center Stage: Philosophy of Design

The scenes and discussion above illustrate the main ideas in our design philosophy which builds on, and puts into practice, the design ideals discussed as background in the first chapter. This design philosophy emphasizes the following points:

- *Cooperative Design.* Users, as well as professional designers, have knowledge and skills that are central to the design of useful computer applications; therefore, design needs to be organized as a cooperative activity between the users and the designers.

- *Family Resemblance.* To allow both groups to contribute effectively and creatively, the design could be based on situations having a family resemblance to the prior work experience of both the users and the designers.

- *Practice.* Designing a computer application and introducing it in a work setting will change the work practice. Yet design must take its starting point in the current practice of the users—an invaluable source for the design of systems that will fit newly emerging work practices.

- *Experiencing the Future.* An effective way of allowing users to employ their knowledge and skills is to simulate future work situations, creating the illusion of actually working with the projected system. In this way the ability of the future computer application to mediate work can be tried out, and changes in the use practice can, to some extent, be predicted.

- *Learning and Transcendence.* Finally, learning is an important ingredient in design processes. The different groups involved learn about the work and background of the others, and confrontation with "outsiders" contributes to the understanding of one's own work practices. In addition, in situations where real or simulated work breaks down, where people's involved action suddenly stops and they start to reflect on their own work, this brings to design an innovative character: The opening of possibilities for new ways of doing things; of transcending the traditional practice of the users and of the designers.

Stage Props: Design as Action

Our theoretical and practical understanding of this design philosophy may be divided into two main lines of thought: *practice* and *power*. We will expand on these to illustrate some of the theoretical underpinnings of our work and to give an understanding of its origins.

Practice

The idea that good design must start from users' practice has been informed by the work of the German philosopher Martin Heidegger, through the interpretations by Terry Winograd and Fernando Flores (1986), and Hubert Dreyfus and Stuart Dreyfus (Dreyfus, 1972; Dreyfus & Dreyfus, 1986). From these writers we have developed ideas about *involved unreflected activity*, as a basic way of being.

To put it simply: When someone sees a bus through a window, they might jump up from the table, grab their briefcase and coat, and run through the house toward the bus stop. They do not reflect on the time left or what they should bring along. They are accustomed to acting in this kind of situation, and, as in most other situations, they act without reflection, adjusting to the specific circumstances, such as whether or not the kitchen door is open. Only if the involved action breaks down, for example if the door is stuck, do they stop and reflect on how to proceed. In this view *detached reflection* is something secondary, taking place only when involved action breaks down.

From these authors we have also learned how tools and other objects are *ready-to-hand* in involved unreflected activity; we are not conscious of their presence. For example, the incinerator plant operator uses the flame and the dials in his work without thinking explicitly about these objects. When using a word processor a person might search for the right words, sometimes worrying about the spelling, but not consciously thinking about different functions, such as entering the characters or deleting a word.

When involved activity breaks down, the objects we use become *present-at-hand*, and we may reflect on them. With a word processor the user might frequently hit the insert key instead of the backspace key. This breakdown makes the person aware of the details of the keyboard operation, causing her to think about possible ways to remedy the situation. For design this example illustrates that only by *use*, not by reflection, can we get to know how the future application will work.

In his book *Work-Oriented Design of Computer Artifacts*, Pelle Ehn (1989) elaborated a complementary way of thinking about different design situations and their relation to use by means of Wittgenstein's notion of language games and family resemblance. According to Wittgenstein's concept of practice, using language is to participate in language games (Wittgenstein, 1953). In discussing how we, in practice, follow (and sometimes break) rules, he asks us to think of games and how they are made up and played. Why games? Because we understand what counts as a game; not because we have an explicit definition, but because we are familiar with other games. In one of the scenes discussed in this chapter, the typographer knows the language game of traditional, computer-based typography. He knows that there are codes in the text. His knowledge has developed over the years from its *family resemblance* with the work of traditional lead-based typography. And it has developed through actual work with this kind of computer application. The language game of presenting and discussing the WYSIWYG (What You See Is What You Get) design, however, is so far removed from what he knows that it doesn't make sense to him. Language games, like other games, are social activities. To be able to play these games we have to learn to follow rules: rules that are socially created and often not explicitly stated. When the typographer talked about codes appearing on the screen, he was really telling the designers that he had not understood the language game of WYSIWYG typography.

In her book *Through the Interface*, Susanne Bødker (1990) applied another theoretical approach, that of Activity Theory, in a discussion of the design of user interfaces. Bødker and other writers in the tradition of Activity Theory draw attention to the *historical* aspects of a shared practice. It is this shared practice that allows people to talk about more than individual skills, knowledge, and judgment. The typographers or incinerator plant operators, for example, are more than "generic" human beings; they share experiences with specific tools, materials, and products. We can view the artifacts that we design as parts of such practices. The need for a new artifact arises out of a certain practice, and the new artifact will, once it is designed and implemented, be reintroduced into that practice, where it will effect changes.

According to Activity Theory, human work and being (activity) is always mediated by artifacts such as tools and language. People do not focus their attention on these artifacts; instead artifacts help focus attention on other objects or subjects. The typographer uses paste and scissors to do paper paste-up, to focus on the newspaper page; the incinerator plant operator uses the camera picture of the flame to control the incineration. These artifacts are present when people are introduced into an activity, but the artifacts are also a product of that activity, and as such they are constantly changed through the activity. Each action in an activity is given meaning through the practice shared with others. For example, writing a specific code in computer-based typography indicates to both the computer and the newspaper workers that the line following this code is a headline. The writing of the code is at the same time based on the physical and social conditions of the situation that the typographer is in. These conditions "trigger" actual, detailed ways of acting. The way of making a headline depends on the technology that the typographer is using, and whether he needs to type some letters or point and select some text.

Artifacts have no meaning in isolation; they are given meaning only through their incorporation into social practice. It is not until they have been incorporated into practice that they can be the basis for thought and reflection. When we design artifacts, they can serve to *mediate* the activity of the users, and at the same time become meaningful tools in the users' practice.

Power

The second line of thought is that of power and its inherent issues of resources and conflicts. A design process where a group of users and designers cooperates does not take place in a vacuum. In most organizations, some groups have more power and better resources than others. Those who have the most power and resources are usually management, not the end-users. To help users get a forum where they can take an active part in design means to set up situations where they can act according to their own rules, and not simply according to the rules of their managers. The weakest groups in particular need the strongest support in formulation of their demands and ideas for the future. There may be conflicting interests among the groups, but as described in Activity Theory these conflicts may be turned into resources for the project, if the situations are set up in appropriate ways. This may mean that designers and users should not just establish one project group of all involved parties, but should work with different groups of people at different times (Ehn & Kyng, 1987; Ehn & Sandberg, 1979; Engeström, 1987). At the same time, the different groups need to be exposed to each other's demands and suggestions, which emphasizes the need for designers as coordinators of the activities. In setting up ways to work with such groups, we can learn from women's research which focuses ways of working cooperatively (Keller, 1985; Greenbaum, 1990; Bødker & Greenbaum, 1988).

In *A Feeling for the Organism*, Evelyn Fox Keller (1983) talks about research methods for "feeling for," or "listening to," the subject of the research. Instead of descending on a research topic with a set of preconceived categories, the idea is to think of oneself as being part of the material. Scientific work in this perspective is living and being involved in the subject matter, or "getting a feeling for the organism." In design activities with resource weak groups it is possible to work in a similar way. Our inspiration from Fox Keller tells us that designers, like research scientists, should not lean too strongly on their own understanding, but should enter the process with an open mind. It also means being involved in the action of design and using methods that support involvement.

Setting the Stage

In the first part of this chapter we discussed some of the theoretical issues involved in design as action, thus setting the stage for Part II of this book. In the second part of this chapter we present some practical ideas and questions which the reader may find useful. We do not, however, present a new step-by-step method. On the contrary, we advise readers to pay attention to the situations in which design is taking place, and to modify, develop, and apply those tools and techniques that seem most appropriate. Our focus is on the *process* aspects of doing design; building on designers' experience while collaborating with the users and learning as we go along.

Because this way of designing is new to us as designers, we may often be in situations where the best path leads away from what is most familiar to us; where our unreflected involved activity is breaking down, and only detached reflection will get us out of the trouble. This obviously creates problems for those who dislike uncertainty. But there is no simple way out—it is a manifestation of the paradox of design: On the one hand, design must be firmly rooted in the work practices of both users and designers; on the other it must confront these practices with their existing shortcomings and introduce new artifacts such as mock-ups, prototypes, or metaphors, and potentially cause breakdowns. This is how we may learn to transcend our own tradition as users *and* as designers. If we stick to our old well-known and "secure" design practices, we as designers end up like those caricatures of users who resist change; an excuse many designers claim for not involving users in the design process.

The next section is about some resources that we find to be necessary requirements for active user participation. Following this we mention a few techniques that we consider helpful in working with users (Kyng, 1988). The chapter ends with a discussion of some critical questions that are often asked in relation to designers' cooperation with users in the design process.

User Controlled Resources

One of the most frequent obstacles to increased user participation in design projects is resources, or rather, the lack of them. Throughout a design project resources must be secured for user participation if there is to be real cooperation. Users who get no reduction in their daily workload cannot be expected to engage themselves in a design project over long periods of time. But resources are not only time and money; they also include education and assistance from a variety of experts.

The right to make decisions about how to spend resources is another crucial aspect of the resource question. Most projects have to follow the rules of the organization to which they belong. In some cases this implies that a project group consisting of designers and users collectively makes the important decisions, as was the case with the UTOPIA project mentioned earlier. In most cases, however, standard managerial procedures will determine the way decisions are made. When a crisis or conflict arises under these circumstances, decisions often run contrary to the wishes of users, and perhaps contrary to those of the designers as well. The result is usually that the users withdraw from the process, either totally, by leaving the design group, or by becoming passive members, losing their initial belief in the possibility of their influencing the design. To give users a better chance of continued participation, we recommend that from the outset the designers argue for a certain amount of resources to be allocated for user initiatives.

Workshops

Design projects are new and unfamiliar to most users, and although the techniques described in this book are intended to facilitate the participation of users without prior design experience, the process of learning can be slow and difficult. Special attention has to be given to the possibilities for users to improve their understanding of the design project and the project's relation to the organization. The users especially must understand their own interests and needs. Future Workshops and other techniques described in the following chapters may contribute to this understanding. Our experience tells us that there is a need for setting up ad hoc workshops with users to provide them with ways to improve their understanding of the design process. In order to limit the effects of unbalanced power and resources in the design process, we suggest that workshops be made up of people from similar levels within an organization.

Where there are real, deep-seated clashes of interest, we recommend that these be dealt with outside the workshops in negotiations between representatives of the groups involved (Ehn & Sandberg, 1979). Conditions for such negotiations could be set up in advance. If there is a union, it should be involved at an early stage. When feedback to management is needed, the designers can talk with the appropriate managers, but only after the participants have been asked; or the participants can talk to management themselves. There may be nothing worse than taking problems from behind the closed doors of a workshop and airing them in public.

To actually get a workshop started, the rules of the game are important. That rules are one of the main ingredients may seem counter-intuitive when we are talking about cooperation and user controlled activities. What we have found, however, is that strict rules break the traditional pattern of communication, often allowing time for more people to speak and interact (see Chapter 8).

Some techniques like going around the room and asking everyone to say something are helpful. This typically results in the presentation of what people might consider to be individual problems of no general interest. But these problems are often shared by many people, leading to new questions related to the use of computers. Thus "individual" problems should be taken seriously, for they may denote larger shared concerns.

Furthermore, even though the practice of the users is the basis for discussions, the designers must use *their* professional practice, in setting up workshops and in pointing out which proposals are in their view feasible. This role of workshop facilitator and technical expert does not imply, however, that the designers should take responsibility for creating consensus.

We think that limited, intensive group activities are useful in the early stages of design. Activities such as workshops that are a little too short, but intensive, are preferred to those where the participants run out of energy and contributions. We find that many groups of users expect daily results, and in lengthy ongoing processes like system design projects, user involvement often decreases when the progress and results are difficult to see. On the other hand it may be unrealistic to expect users and designers to come up with daily results. Short, intensive activities where participants work actively together within a limited time frame may help, but such activities require commitment. This means trying to avoid interruptions such as phone calls or meetings. It also means, as we emphasized earlier, that resources should be made available so that the participants get time off from some of their usual tasks in order to participate in design.

Workplace Visits

We have discussed getting users started in the process of looking at their own situation and identifying the bottlenecks, difficulties, and other areas that they want to change. The second half of this story is about developing ideas for new uses of technology. Often the existing technical solutions at a specific workplace severely limit the creativity of the users. Activities are needed to stimulate the "technological fantasy" of the users. Several of the techniques described in the following chapters may be used to achieve this, but more modest ways have also proven useful. As a step toward understanding different possibilities, a design group may look at some existing computer applications. Material describing cases of similar or more advanced use is often helpful for the users; visits to other workplaces offer such concrete examples.

Visiting workplaces is a simple and powerful way of getting to understand that a broad spectrum of possibilities exists. These workplaces should preferably be in the same line of business or in important respects similar to those of the users and, of course, they should have noticeable technical solutions. A team designing computer applications for newspapers, for example, may visit different printing offices and studios producing advanced moviegraphics. Seeing the technology at work, and discussing it with its users, immediately allows the visitors to have access to experiences from use. In some cases it is possible to arrange for the visitors to work with a system for a period of time, thus making the experiences from use partially their own.

Ready to Act?

We end this chapter by pointing to and discussing some of the questions that designers often have when talking about cooperative design. We present them here to help the readers phrase their own questions as they read the second part of the book. We return to these issues and other questions in Chapter 13.

What happens when major conflicts and power struggles arise? They are probably going to be there anyway, so why wait until a system is developed to see the sparks fly when new procedures bump into older, established ways of doing things? Getting to know more about the involved groups and helping them to articulate their concerns may help the designers understand that there are fundamental conflicts between groups in an organization (i. e., management vs. labor, between different departments, and between different groups of workers), and thus understand why apparently irrational arguments sometimes develop. In many instances, designers cannot solve these conflicts, but workshop methods, as they are described in this chapter and the next, can help the involved groups better articulate their own needs, and the designers can help the groups discover how they may act to change their own situation.

How should managers be involved? Generally one needs to be careful about how design groups are put together. Power relations among people in a work group often prevent the weakest parties from contributing anything to the group. Furthermore, managers are often more used to "taking the floor" and expressing themselves in various settings. Groups of users at similar work levels seem to be the best place to start for certain issues. Of course this can include groups for managers. For some issues it may be necessary to put together groups with differing powers and resources. In these cases support groups for the participants from the shop floor can be organized to let the weaker groups help themselves.

Isn't design as action a slow way of getting things done? What about deadlines? Experience shows that project estimation is difficult and that many projects do not stick to their deadlines. This problem is not solved by the approaches used in this book. In fact it can even be argued that stimulating creativity among users implies a higher uncertainty for project estimation, because designers cannot anticipate what ideas might come up during the early design activities. Generally, our approach should be thought of as an iterative one, where the introduction of the new computer application in the work setting may cause unforeseen changes in work, in turn leading to demands for new or changed computer applications. But we also believe that our techniques of involving users in projects will improve the quality of the developed products and help anticipate some of the more expensive design errors that occur in traditional system development. Thus, a project inspired by the techniques in this book will not automatically be more expensive than a traditional project. Indeed, ours is a comparatively "front ended" approach, where user-designer interaction is emphasized from the start, and therefore may alleviate later problems.

The role of designers becomes rather complicated in the cooperative design process. Designers are in charge of the project, they are responsible for getting the work going, and they must be able to act as facilitators of workshops and similar events, and, in general, as resources for the groups. And designers are the ones who make sure that all the technical details are in place.

One of the issues confronted when we tear down the barriers of traditional design is that the role of the system developer shifts from that of project manager to project facilitator. Cooperative design, which by definition means empowering users to fuller participation and cooperation, breaks down the old rules of the game. Within traditional system development each step, from feasibility study through implementation, is supposed to be controlled by system developers through discrete procedures, and marked by clearcut milestones and exit criteria. In our view, the traditional system approach may have only made it easier for designers to *look* like things were under control. Like good actors, system developers played the role of project managers, but the performance sometimes unraveled in the last act when the system was actually placed in use. The cooperative design approach *begins* by trying to create an environment where users and designers can actively view the use situation. Designers shouldn't have to wait until the final act to know whether or not the system will fit the practice of the users. This changes the designer's role, we believe, by opening up more space for users to act. It also raises questions about how designers can learn to play more of a facilitator role. As Mike Cooley points out in the opening quote of this chapter, the term design has been used since around the sixteenth century to imply a separation of the people who do design from those who do less conceptual work. We think it's about time for that to begin to change.

Acknowledgments

We want to thank Pelle Ehn, Jonathan Grudin and Lucy Suchman for their contributions to this chapter.

References

Boehm, B. (1988). A spiral model of software development and enhancement. *Computer, 21* (5), 61-72.

Bødker, S. (1990). *Through the interface—a human activity approach to user interface design.* Hillsdale, NJ: Lawrence Erlbaum Associates.

Bødker, S., Ehn, P., Kammersgaard, J., Kyng, M., & Sundblad, Y. (1987). A utopian experience. In G. Bjerknes, P. Ehn, & M. Kyng (Eds.), *Computers and democracy—a Scandinavian challenge* (pp. 251-278). Aldershot, UK: Avebury.

Bødker, S. & Greenbaum, J. (1988). A feeling for system development. In K. Tijden, M. Jennings, & U. Wagner (Eds.), *Women, work and computerization: Forming new alliances, Proceedings of the IFIP TC 9/WG 9.1.* Amsterdam: North-Holland.

Cooley, M. (1988). From Brunelleschi to CAD-CAM. In J. Thackara (Ed.), *Design after Modernism—Beyond the Object.* New York: Thames and Hudson.

Dreyfus, H. (1972). *What computers can't do.* New York: Harper & Row.

Dreyfus, H. & Dreyfus, S. (1986). *Mind over machine—the power of human intuition and expertise in the era of the computer.* Oxford: Basil Blackwell.

Ehn, P. (1989). *Work-oriented design of computer artifacts* (2nd ed.). Hillsdale, NJ: Lawrence Erlbaum Associates.

Ehn, P., & Kyng, M. (1987). The collective resource approach to system design. In G. Bjerknes, P. Ehn, & M. Kyng (Eds.),

Computers and democracy—a Scandinavian challenge (pp. 17-57). Aldershot, UK: Avebury.

Ehn, P. & Sandberg, Å. (1979). God utredning [Good investigation]. In Å. Sandberg (Ed.), *Utredning och förändring i förvaltningen.* Stockholm: Liber.

Engeström, Y. (1987). *Learning by expanding.* Helsinki: Orienta-Konsultit.

Greenbaum, J. (1990). The head and the heart. *Computers and Society, 20* (2), 9-16.

Keller, E. F. (1985). *Reflections on gender and science.* Yale: University Press.

Keller, E. F. (1983). *A feeling for the organism: The life and work of Barbara McClintock.* New York: Freeman.

Kyng, M. (1988). Designing for a dollar a day. *Office, Technology and People 4* (2).

Sundblad, Y. (Ed.). (1987). *Quality and interaction in computer-aided graphic design.* (Utopia Report No. 15). Stockholm: Arbetslivscentrum.

Winograd, T. & Flores, F. (1986). *Understanding computers and cognition—A new foundation for design.* Norwood, NJ: Ablex.

Wittgenstein, L. (1953). *Philosophical investigations.* Oxford: Oxford University Press.

8

Generating Visions: Future Workshops and Metaphorical Design

Finn Kensing and Kim Halskov Madsen

In this chapter we are especially concerned with supporting the generation of visions for the future use of computers. A critical problem here seems to be that user-designer cooperation is poorly supported by current methods. Empirical research shows that while a lot of resources might be used on interviewing users about their current work, few or no resources are used in helping users and designers generate alternative ideas about how they would like their work situations to be in the future (Andersen, Kensing, Lundin, Mathiassen, Munk-Madsen, Rasbech, & Sørgaard, 1990). Kalle Lyytinen has surveyed the relevant literature and presents eleven problem classes regarding development and use of computers in organizations (Lyytinen, 1986). A few of those classes which relate to the theme of this chapter are:

- development activities pay only slim attention to changes of job-content, autonomy, work-load and so forth introduced by the new system ....
- system goals are ambiguous and often poorly defined, and if defined too narrow, because they concern mostly technical and economic issues ....
- system goals motivate mainly systems-analysts and management and not the end-users. The result of this is that the devel-

opment process seldom solves the "right" user-problems, because these are insufficient incentives for the user to participate and contribute his know-how ....

- process is a specialist driven activity and it tends to focus on the average user instead of a unique individual and his needs ....

- methods and tools employed are orientated toward improving the work of the systems-analyst and programmers; they do not help much users to take part in the process and build systems they really need ....

- in general conceptual problems result in solving the wrong problems instead of the right ones ....

- negative attitudes or reactions toward the built system which find their expression in a multitude of ways from Luddism and sabotage to a lack of motivation to use the information system as appropriately as possible. (pp. 4-9)

These problems relate to cooperation between designers and users when generating visions, or, as Kalle Lyytinen says, when defining system goals. We advocate that the process of creating visions about future work situations should be explicitly supported and derive from an interplay between the competence of users and designers. For that purpose, the toolbox of most designers would benefit from being supplemented with approaches less formal than those currently used.

We suggest Future Workshops and metaphorical design as examples of new approaches. In this chapter we demonstrate how the combination of Future Workshops as an organizational frame and metaphor as a linguistic tool can stimulate creative visions of the future use of computers in organizations. In the conclusion we address how this approach to design helps designers by taking into account Lyytinen's problem classes.

Background

Robert Jungk and Norbert Müllert have developed a technique called Future Workshops (Jungk & Müllert, 1987). The technique was originally developed for citizen groups with limited resources who wanted a say in the decision making processes of public planning authorities (town planning, environmental projection, energy crisis, etc.). Finn Kensing has proposed its use in system development (Kensing, 1987). It is a technique meant to shed light on a common problematic situation, to generate visions about the future, and to discuss how these visions can be realized. Those participating should share the same problematic situation, they should share a desire to change the situation according to their visions, and they should share a set of means for that change. The technique is described briefly and its use is shown in the scenario presented in this chapter. The reader wanting to practice the technique is strongly encouraged to read Jungk and Müllert's book, which has been translated into several languages.

As a way to broaden the perspective of the participants, we encourage the facilitators to intervene at the content level by introducing metaphors. The use of metaphors is helpful if the participants get stuck or develop their critique or visions in too narrow a way. Generally, a Future Workshop is run by one or two facilitators, with no more than twenty participants. The facilitators attempt to ensure an equal distribution of speaking time and they should also ensure that all participants can follow the discussion, by letting the participants write their ideas as short statements on wall charts (a large sheet of paper taped to the wall). The form and content of the phases are described later in the scenario.

A Future Workshop is divided into three phases: the Critique, the Fantasy, and the Implementation phase. Essentially the Critique phase is designed to draw out specific issues about current work practice; the Fantasy phase allows participants the freedom to imagine "what if" the workplace could be different; and the Implementation phase focuses on what resources would be needed to make realistic changes. These phases are surrounded by preparation and follow-up periods.

We find that metaphors stimulate seeing things in new ways. They are perhaps most well-known from poetry, for example, "Thou blind fool, love, what dost thou to my eyes," from Shakespeare's Sonnet 137. In this kind of metaphorical personification, attributes normally assigned to human beings are assigned to an abstract concept like love. The essence of metaphors is to talk about one thing in terms of another, the two things being different in some way (Lakoff & Johnson, 1980). But metaphors are not reserved for poets—they pervade our entire life. For example at a library, workers understood the computer system in terms of a physical space: "I am in the circulation control system" and "Now I have to go to the ALIS-base," (Andersen & Madsen, 1988). Hence, metaphors, in contrast to formal system description tools, are a natural part of everyday language.

Donald Schön has discussed the relevance and usefulness of metaphors in the area of design in its broad sense (Schön, 1979). Giovan Francesco Lanzara has pointed out their specific relevance to the design of computer systems (Lanzara, 1983), and Kim Halskov Madsen has given concrete guidelines for the use of metaphors in the design of computer applications (Madsen, 1989). As we will see, metaphors can be used as a tool for reflection as well as for action; thus, metaphors support workers' reflection on their current view of their own work and stimulate their visions of alternative future ways to work.

In the following we describe a combination of Future Workshops and metaphorical design, currently being developed through experimentation. For several years we have arranged Future Workshops, developed metaphorical design in the context of system development, and carried out the "initial actions" mentioned below. We present our approach to design as a scenario, in the sense that the actual course of actions described has not taken place. Instead, the scenario is constructed from our experiences as consultants and teachers for various groups of users and designers in a variety of cases. The setting, the characters and the context of the scenario are taken from a technology assessment project at the Danish research libraries (Etzerodt & Madsen, 1985; Etzerodt & Madsen, 1988). Metaphorical design was applied in the project; however, we did not conduct a Future Workshop in that project. In the scenario of this chapter we have incorporated experiences from Future Workshops as well as experiences from applying metaphorical design in other cases. The idea behind merging our experiences into one scenario is to help the reader imagine a blending of these techniques.

The Overall Scenario

The Danish public library system consists of a number of county libraries, each with a smaller number of branches. Each library is made up of two departments, the Accession department and the Circulation department. In the Accession department, books are chosen at book selection meetings; later the books are ordered, received, and registered. In the Circulation department the borrowers, perhaps assisted by the librarians, choose books to read. At the counter, clerical workers handle lending and returning of books and send out overdue notices.

Although each county library and its associated branches are independent, the libraries have a long tradition of close cooperation. A number of central institutions provide various services to the libraries, such as bookbinding and materials for book selection meetings.

Twelve years ago these institutions, together with the main software supplier for public institutions (DMK), initiated a large development project called The Library Data Project. As a result, DMK is in a position to deliver computer support for information retrieval, circulation control, and cataloguing. The core of these systems is a central database, called BASIS, which is shared by all libraries. At the library where we were consulting, the information retrieval system had been in use for a couple of years. It was fairly advanced and required extensive training to use. Borrowers couldn't use the system on their own.

The circulation control system and cataloguing systems were under consideration by the chief librarian. She felt a need to demonstrate efficiency in order to deal with coming government spending cuts, and considered computerization as a way of increasing efficiency. In this way time would be made available for improving the service. For example, by computerizing cataloguing, librarians from the Accession department could be moved to the Circulation department where they could assist borrowers.

The librarians had a fairly hostile attitude toward the use of computers. They believed that a computer couldn't do anything better than they could do it themselves. The clerical workers were ready to discuss efficiency, but only together with improved quality of service. They feared that computerization of the library would eventually lead to a loss of jobs. None of the groups believed in the chief librarian's idea of moving librarians to the Circulations department. They felt that it was unrealistic at a time when the libraries were facing government cuts.

Although management, DMK, and the government originally argued that circulation control should have top priority, the library staff had managed to argue for a need for improved quality of service through the information retrieval systems. But now the staff realized that the quality of service had not improved radically. Hence, the library staff was unsatisfied with the concept of the DMK systems, although they couldn't find alternatives. We will imagine that they have asked us, as outside consultants, to assist in generating alternative suggestions challenging the concept of DMK. The following is an outline of what could have happened.

Initial Actions of the Scenario

After a few meetings with the librarian who had contacted us, we advocated the formation of a project group consisting of three librarians, three clerical workers, and the two of us. This group promised to deliver a proposal of future use of computers at the library in four months.

The group started out by developing a common background against which the work was to be done. We worked at the library as "apprentices" for one week. During that week we paid special attention to the patterns of interaction among the staff and between the staff and borrowers, to the work language, and to the physical surroundings. Because only the librarians had computers and because they had been using only the information retrieval system, we found it important to stimulate the awareness and the fantasy of the staff by demonstrating hardware and software and by arranging visits to other workplaces. We visited libraries already using the full-scale DMK system, and similar workplaces such as a store, a museum, and a local community and activity center. At the various places we arranged meetings with the staff to discuss their experiences regarding development and use of computers in their work. The aim of these activities, however, was not to "sell" the idea of using computers, but to generate ideas about the implication of possible uses of computer.

Our visits to workplaces other than libraries were particularly important because we got ideas for metaphors that stimulated alternative views of how computers could be used at a library. We created the initial metaphors without involving the staff from the library because we have found that it is hard for people to come up with metaphors about their own work. But people learn by example—it is much easier to create additional metaphors after hearing two or three plausible examples.

The common background established by these initial actions could have been supplemented by an analysis of the organizational culture as proposed by Keld Bødker and Jesper Strandgaard Pedersen in Chapter 6, or by an analysis of the work language as proposed by Berit Holmqvist and Peter Bøgh Andersen in Chapter 5.

In addition to forming the picture of the situation at the public libraries, our analysis led to the idea of three different metaphors for how to interpret what goes on in a library: a *warehouse*, a *store* and a *meeting place*. The metaphors often grew out of one or two concepts that were "translated" into concepts normally used about a library. For instance, the goods-books relation led to a warehouse metaphor and the customer-borrower relation led to a store metaphor. Seen as a warehouse, the library is a place where books are stored. Seen as a store, focus is on the service toward the borrow-ers. Seen as a meeting place, the focus is on relations among bor-rowers and staff.

For the library staff the most important effect of the initial actions of the group was the stimulation of their technical and social fantasy. For example, we had visited workplaces where high-resolution work stations were used as well as workplaces where just terminals were used. We had visited workplaces with a very sharp division of labor between librarians and clerical workers, as well as others with a more flexible division of labor. The staff learned about some of the possibilities and hindrances regarding the use of computers in their work; indeed we noticed that "technology and work" appeared in daily gossip.

Metaphorical Design at the Future Workshop

The project group decided to set up a one-day Future Workshop. The project group worked out an invitation announcing the theme of the workshop: "Computer technology on our terms." In addition, the invitation briefly presented the overall idea, the phases of the workshop, and a program for the day. The invitation was sent to all employees at the county library and its seven branches. Though the main conflict had been between local government and DMK on the one hand, and library employees, regardless of rank, on the other, the employees decided not to invite the library heads. The main ar-gument for this was that the employees did not share the chief librarians' ideas about how to increase efficiency.

The project group then found a suitable place and provided the needed materials, which included tape, markers, and large sheets of paper to be used as wall charts. However, the actual setting up of the room was done in cooperation with the participants in order to create a relaxed atmosphere and to emphasize that it was not just another meeting they were attending.

The Critique Phase

At the beginning of the Future Workshop the facilitators introduced the technique to the participants, and the plan for the day was discussed and adjusted. Basically the Critique phase is like a struc-tured brain-storming that focuses on current problems at work. The Critique phase opened with criticism of DMK's plans as well as criticism of the staff's own work practices. The facilitators ex-plained that as the participants suggested their critiques, they were to be formulated as short statements and written as a few keywords on the wall chart. Speaking time was restricted to 30 seconds to make it easier for all participants to speak (Jungk & Müllert, 1987). The facilitator also explained that participants did not need to defend or offer arguments for their ideas, thus enabling less verbal workers to jump into the process.

At one point the facilitators intervened and pointed out that the library staff talked about the library as if it was a warehouse; they suggested that the library alternatively could be seen as a store. For example, seeing the library as a warehouse made people think about the way books were kept there, whereas seeing the library as a store brought the focus more toward serving "customers" in the library. Often, the metaphors led to the critique being formulated as a caricature: "The library is like a supermarket"; thus, the critique was stated more clearly. In the discussions the specific metaphors were used, but the concept of a metaphor was not. These metaphors are discussed more fully in the next section on the Fantasy phase.

Here are some of the short statements as they appeared on the wall chart:

```
too slow throughput
poor knowledge about ordered books
no support for loss of books
all we do is inventory control
the library is like a supermarket
DMK has no support for revision
never talk to the borrowers
efficiency instead of service
DMK favours centralization
fear of centralization
the demarcation is the obstacle
the library is just a museum today
librarians are just attendants
we buy computer equipment instead of books
fear of loss of jobs
we only provide self-service
poor marketing
no computer support for better service
```

Figure 1. Statements from the Critique phase.

To get an impression of the general drift of the critique, the participants together with the facilitators grouped the short statements under the following headings:

• The library as a warehouse

- Relation to the borrowers
- The organization of the library
- The library as a store
- The role of DMK

The participants were divided into small groups of four or five people. Each group chose a set of short statements and through discussion reformulated these into a concise critique of the current use of technology and the plans of DMK. The discussions in the groups were outside the control of the facilitators. In the subsequent plenary session the critiques of the groups were presented and discussed.

The Fantasy Phase

To stimulate the imagination of the participants, the Fantasy phase started with two warm-up activities. One was simply to invert the short statements from the Critique phase into positive statements. The other was to draw pictures of the library as the participants would like it to be in five years. The drawings were hung and a brainstorming session was started. Again, long speeches were prohibited, and the facilitators cut off "killer formulations" like: "This is completely unrealistic!" No statement about future working situations and future computer applications was considered too extreme. Again the participants were encouraged to write short statements on the wall chart, some of which are:

> tear down the walls
> an electronic bulletin board
> arrange small reading groups
> a review database
> overview of most popular books
> re-use of other peoples synonym lists
> better support for self-service borrowers
> direct contact with the authors
> create connections among borrows from
> different libraries
> access to the library from home
> no division of labour

Figure 2. Statements from the Fantasy phase.

As in the former phase, we wanted an impression of the general drift of the statements. But in this case, we employed a ranking system where each of the participants had five votes to cast on the statements they favored. The seven short statements getting the highest score were summed up in a "utopian outline" under the heading "The library seen as a meeting place." In the small groups, the outline was discussed and further developed, while still ignoring possible drawbacks, which were to be discussed in the next phase.

But before the group discussions, the participants were introduced to the idea of metaphorical design in order to stimulate their talk (Madsen, 1989).

We used the warehouse metaphor as an example. We pointed out to the participants that when we had talked about the library as though it were a warehouse, we didn't mean the library literally was a warehouse. The point was that certain aspects of a library could be highlighted by comparing it with a warehouse. We could do that more systematically by comparing aspects of the warehouse with similar aspects of the library, such as "stock in trade" with "book stock," "orders" with "requisitions," "stock taking" with "revision," "purchase of goods" with "accession," "delivery of goods" with "lending of books." More similarities like these could be generated by focusing on the characteristic aspects of a warehouse and by identifying similar aspects of a library. The intention was not only to be aware of the similarities between a warehouse and a library, but just as importantly to clarify the differences between the two. "Stock in trade" is quite similar to "book stock," whereas "purchase of goods" is rather different from "accession," because almost any purchase of a book is a unique case.

We encouraged the participants to be aware of the part of the library that was highlighted, and also the part that was hidden, by the warehouse metaphor or warehouse view, as we also called it. Those parts of the library that were left out or hidden would be likely candidates for other useful metaphors. In the case of the warehouse metaphor, the borrowers were left out. Borrowers could give rise to a store metaphor, because a store is like a warehouse but with customers.

We pointed out that, seen as a warehouse, the most important tasks of the library are to find the goods quickly, to have precise knowledge of the stock, and to avoid loss. A good computer system is one that makes it possible to keep track of the goods, and it seems that the existing computer systems offered by DMK—the accession system, the information retrieval system, and the circulation control system—are usable for this purpose.

The Implementation Phase

The Implementation phase started by having each group present their version of a utopian outline. Inspired by a meeting place metaphor, one of the groups had considered how the library could be seen as a place for conversations about books and other book-related subjects. They had noted that the conversations are not solely about books, but also about reviews of books relevant to cultural events in the city. Conversations take place among the borrowers, among the staff, and between the staff and the borrowers. Among the suggestions for computer systems was a new book catalogue and an electronic bulletin board to support conversations among the borrowers.

During the discussion of the outline it became clear that when the library is seen as a meeting place, the most important task of the library is to create better conditions for communication. In this context a good computer system is one that can be used as a medium for conversations, or one that can establish contact between people and thereby create possibilities for conversations.

Later in the Implementation phase, the outlines were evaluated in a plenary session to see whether it was possible to realize them under current conditions. In addition, it was discussed whether it was necessary or possible to establish new conditions under which the utopian outline could be realized. The discussion was carried out briefly in the plenary session and more thoroughly in smaller groups, where suggestions were worked out for how the visions could be brought about as well. The suggestions were discussed and coordinated in a plenary session, where they were turned into a common strategy for the library staff.

The Future Workshop closed with the participants making a detailed plan for how the first steps in the strategy should be taken. The plan consisted of a list of tasks to be carried out within a given period and the names of the people who had signed up for each of them. The following is part of the plan dealing with a scheme for realizing the meeting place metaphor, including the idea of an electronic bulletin board.

- Susan, Peter, and Albert. Within two weeks: Contact the consultants of DMK to interest them in developing a prototype of the bulletin board.
- Lars, Hans, and Lone. Within two weeks: Contact some of the other county libraries to get their support or acceptance of DMK giving high priority to the development of a prototype.
- Jesper and Lea. Start next week: Develop a set of criteria, based on the short statements from the Critique and the Fantasy Phases,

for evaluation of the prototype. Arrange a meeting for discussion of the criteria.

Workshops like the one described have often formed a good basis for further work of project groups. Plans for specific actions are an important outcome of future workshops, and the wall charts can provide crucial documentation of the workshop itself. A future workshop is actually an ongoing activity or, as Jungk and Müllert (1987) point out, if the group starts implementing the plan, a "permanent" workshop arises and the future workshop techniques may be reapplied when obstacles occur. But, of course, other techniques are also needed and we suggest that readers turn their attention to the other chapters in Part II.

Discussion

We have suggested the idea of using Future Workshops and metaphorical design as an approach to generating visions, a neglected part of system design. How does this approach help designers handle the examples we selected from Lyytinen's survey of problems mentioned in the literature (Lyytinen, 1986)?

It is our experience that Future Workshops and metaphorical design helps define system goals by focusing on issues related to how to get the job done, rather than on technical and economic issues. The approach makes it possible for the users to take an active part in contributing their knowledge, thereby helping the "right" problems to be solved.

Future Workshops and metaphorical design, though facilitated by the designers, represent techniques that are more user-driven than traditional methods. The orientation is toward helping users take part in the design process. This is made possible mainly by communicating in everyday language and by focusing on the actual users and their needs, rather than on the average user.

Future Workshops and metaphorical design provide a framework for paying attention to changes in the working environment and in the organization. Playing with metaphors in the Critique phase and in the Fantasy phase makes it easy for the users to express their relevant likes and dislikes.

In our experience, being facilitators of Future Workshops and playing with metaphors gives us an idea of the "soul" of the organization in question. Like the ideas about workplace culture discussed in Chapter 6, we find that this helps us better understand how users envision their work environment. This is crucial for our further conceptualization of the system, and it is a kind of understanding that we have not been able to acquire using traditional methods.

Future Workshops and metaphorical design do not, of course, preclude negative attitudes. However, since the very idea is to make it possible for users to develop their ideas of more desirable systems based on a critique of the current ones, it decreases the chances of finding negative reactions later in the development process.

In the scenario described we have assumed that the workshop went smoothly, in order to illustrate the main points of the approach. But we have experienced some practical problems applying this approach to design; for instance, time pressure during the Future Workshops; ensuring that the plans are actually carried out; selecting a coherent group of participants; and ensuring that the facilitators inspire without manipulating. How to handle such problems is, of course, crucial to the success of the approach. However, we see such problems as not being limited to our approach, as all workplace centered methods require a lot of time. Instead, we hope to have illustrated how the approach applied here seriously takes into consideration the competence and knowledge of the users about their workplace. The whole idea of Future Workshops and metaphorical design is to allow people to enlighten their common problem situation, to generate visions for the future, and to discuss how these visions can be realized. The aim is to support users playing an active role in the design process.

Acknowledgments

Thanks to Pelle Ehn, Elin Rønby Pedersen, Dan Sjögren, Randy Trigg, Terry Winograd and the Editors for many helpful comments on earlier drafts.

References

Andersen, N. E., Kensing, F., Lundin, J., Mathiassen, L., Munk-Madsen, A., Rasbech, M., & Sørgaard, P. (1990). *Professional systems development: experience, ideas and action*. London: Prentice-Hall.

Andersen, P. B. & Madsen, K. H. (1988). *Design and professional languages*. (DAIMI PB-244). Aarhus, Denmark: Aarhus University, Computer Science Department.

Etzerodt, P. & Madsen, K. H. (1985). Methodological choices when evolving user's understanding of computers. In M. Lassen & L. Mathiassen (Eds.), *Report of the Eighth Scandinavian Research Seminar on Systemeering* (pp. 88-99). Aarhus, Denmark: Aarhus University.

Etzerodt, P. & Madsen, K. H. (1988). Information systems assessment as a learning process. In N. Bjørn-Andersen & G. B. Davis (Eds.), *Information systems assessment: issues and challenges* (pp. 333-350). Amsterdam: North-Holland.

Jungk, R. & Müllert, N. (1987). *Future workshops: How to create desirable futures*. London: Institute for Social Inventions.

Kensing, F. (1987). Generation of visions in systems development. In P. Docherty, K. Fuchs-Kittowski, P. Kolm, & L. Mathiassen (Eds.), *Systems design for human and productivity— Participation and beyond* (pp. 285-301). Amsterdam: North-Holland.

Madsen, K. H. (1989). Breakthrough by breakdown: Metaphors and structured domains: In H. Klein & K. Kumar (Eds.), *Systems development for human progress* (pp. 41-55). Amsterdam: North-Holland.

Lakoff, G. & Johnson, M. (1980). *Metaphors we live by*. Chicago: University of Chicago Press.

Lanzara, G. F. (1983). The design process: Games, frames and metaphors. In U. Briefs, C. Ciborra, & L. Schneider (Eds.), *Systems design for, with, and by the users* (pp. 29-40). Amsterdam: North-Holland.

Lyytinen, K. (1986). *Information systems development as social action: Framework and critical implications*. Jyväskylä, Finland: Jyväskylä Studies in Computer Science, Economics & Statistics.

Schön, D. (1979). Generative metaphor: A perspective on problem setting in social policy. In A. Ortony (Ed.), *Metaphor and thought* (pp. 254-283). Cambridge, UK: Cambridge University Press.

9

Cardboard Computers: Mocking-it-up or Hands-on the Future

Pelle Ehn and Morten Kyng

Mocking-it-up.

This picture shows some artifacts we have used in designing the future of computer-supported newspaper production in the UTOPIA project. There are paper sheets on the wall, slide projectors, screens, racks, chip boards, some chairs, and a cardboard box. However, something is missing. No, it is not computers, but the empty chairs certainly have to be occupied by future users. In our view, artifacts, computers as well as other tools, should be understood via the human use of them.

The users who would be envisioning their future work situation in the design game above are typographers and journalists working with editing, layout, and page make up. The relationship between these two groups has always been a bit tense. Journalists (assistant editors) who work with editing and layout of text and pictures are responsible for the quality of the content of the product (the readability); typographers (make up staff) who work with page make up are responsible for the quality of the form of the product (the legibility). However, the border between the two responsibilities is far from clear.

Page make up work using lead technology.

In the good old composing days journalists worked in the newsroom and typographers worked in the composing room. The editor sent a layout sketch to the composing room and the make up staff returned a "proof" to the newsroom before sending the page to be printed. Not a perfect process, but it worked well. With paper paste up technology it became more difficult to make proofs. It was too expensive and took too much time to run a proof on the photo typesetter. Hence, the assistant editors began to hang around in the composing room controlling the work of the make up staff. Not surprisingly, the typographers were not too happy about this arrangement.

With the introduction of computer-based layout and page make up in the late 1970s, the relations between the "rucksacks" (as the typographers called the journalists in the composing room) and the make up staff got even worse. Now the work was literally taken away from many typographers, since the equipment was placed in the newsroom and was operated by the assistant editors. However, aside from the personal misfortunes of introducing this new technology, the solution was far from optimal in terms of typographic quality.

The design question we were facing in the UTOPIA project was the following: Are there technical and organizational design alternatives that support peaceful and creative coexistence between typographers and journalists, where both readability and legibility of the product could be enhanced?

Now take a closer look at the cardboard box at the right in the first picture. On the front is written "desktop laser printer;" that is all there is. It is a *mock-up*. The box is empty, its functionality is zero. Still, it works very well in the design game of envisioning the future work of assistant editors and make up staff. It is a suggestion to the participating users that an inexpensive computer-based proof machine could be part of the solution. With the help of new tecnnology, the old proof machine can be reinvented and enhanced.

A mock-up of a laser printer "reinventing" the old proof machine.

The journalist makes a layout sketch, sends it to the typographer, and the typographer works on the page make up. Whenever he is in doubt or has suggestions for alternatives, he sends a proof via the desktop laser printer to the journalist, who marks with a few pen strokes how he wants the page to look, and sends the proof back to the typographer, who completes the page. Both can concentrate on what they are best at: the assistant editor on journalistic quality and the make up person on typographic quality. And why should they not sit in the same room and talk to each other?

The mock-ups in the pictures were made and used in 1982. At that time desktop laser printers only existed in the advanced research laboratories, and certainly typographers and journalists had never heard of them. To them the idea of a cheap laser printer was "unreal." It was our responsibility as professional designers to be aware of such future possibilities and to suggest them to the users. It was also our role to suggest this technical and organizational solution in such a way that the users could experience and envision what it would mean in their practical work, before too much time, money, and development work were invested. Hence, the design game with the mock-up laser printer.

In this chapter we will show some prototypical examples of mock-ups. We will discuss how and why they are useful in participatory design. Our examples will range from "cardboard computers" to "computer mock-ups" hinting at the pros and cons of less and more advanced artifacts for envisionment of future use. Finally, the use of mock-ups is put into the perspective of other activities going on in participatory design.

Why Mock It Up?

What we suggest in this chapter is that design artifacts such as mock-ups can be most useful in early stages of the design process. They encourage active user involvement, unlike traditional specification documents. For better or worse, they actually help users and designers transcend the borders of reality and imagine the impossible.

But why do mock-ups work despite their low functionality and the fact that they only are a kind of *simulacrum*? Some of the obvious answers include:

- they encourage *"hands-on experience,"* hence user involvement beyond the detached reflection that traditional systems descriptions allow;
- they are *understandable*, hence there is no confusion between the simulation and the "real thing," and everybody has the competence to modify them;

- they are *cheap*, hence many experiments can be conducted without big investments in equipment, commitment, time, and other resources; and last but not least,
- they are *fun* to work with.

We Did Not Make It Up

Certainly we did not invent the idea of using mock-ups. Kids have always been good at playing with mock-ups like dolls, cars, etc. It is hard to imagine human life without these kinds of games.

However, the use of mock-ups can be most seductive. Think of computer exhibitions. What looks like a running system is often not the final system, nor even a prototype, but simply a video tape or a programmed slide show. Good envisionment of a future product; however, more than one manufacturer has passed the border between concerned marketing envisionment and deliberate manipulation.

Our way of using mock-ups has—in the terms introduced in Chapter 7—a family resemblance to both children's play and envisionment at exhibitions, but the most important inspiration comes from industrial designers. They have been using mock-ups professionally for decades. In particular, they have been successful in using mock-ups in ergonomic design.

Mock-up or the real system? An advertisement for the TIPS page make up system which was based on UTOPIA specifications. When the ad was published no "real" system existed.

One example is the use of ergonomic rigs. This is a mock-up environment in which designers and users together can build mock-ups of, for example, a future work station. Typically there will be support for rapid and cheap mocking-up of ergonomic aspects of appropriate tables, chairs, monitors, etc. Several alternatives can be designed and the users can get hands-on experience. Later, the designers can elaborate the mock-ups as in the following picture, where a future reception workstation has been envisioned.

Industrial design mock-up.

But it is not necessary to be a professional industrial designer to make useful mock-ups. The next picture is from a newsletter published by one of the clerical worker unions in Sweden. It shows a mock-up of a proposal for a new computer-controlled parcel sorting workstation. Originally, the local union was presented with only the technical specifications of the new proposal. However, on the basis of the drawings, the workers were unable to judge the quality of the proposal with respect to the effectiveness of work procedures and physical strain. They then spent a few thousand dollars to build the full scale mock-up. Using this they were able to simulate the future work: the flow of parcels, the tasks of each operator, including work load, and the possibility of supporting each other when bottlenecks occurred. The simulations resulted in several improvements, including suggestions for reducing physical strain and new ways of cooperating.

En lokal fackklubb förbereder sig för ny teknik:
— Ritningarna begriper vi inte
Vi gör attrapper och provar

Sort machine mock-up. The headline reads: "We did not understand the blueprints, so we made our own mock-ups."

The use of mock-ups described here resembles the way industrial designers use them. However, our focus is on setting up design games for envisionment of the future work process. In contrast to industrial designers, we focus more on the hardware and software functionality of the future artifacts and less on the ergonomic aspects. Industrial designers often make very elaborate aesthetic and ergonomic designs of keyboards, but the display is black, and no functionality is simulated or mocked-up. If these different capabilities could meet in a participative design effort, an even more realistic simulacrum could be created. If the future users also actively participate in the design, the mock-ups may be truly useful and a proper move toward a changed reality. But are mock-ups really professional design artifacts? Yes, they are. In arguing this point, we will get a bit more philosophical, but we will also look at the theory in some practical examples.

Language games

We are guided by the concept of 'What a picture describes is determined by its use.' This is shocking statement for those of us who were brought up in a natural science tradition where a system description normally is understood as a kind of mirror image of reality. Nevertheless, this is a position at the heart of Ludwig Wittgenstein's *Philosophical Investigations* (1953). Wittgenstein was aware of this challenge. As a philosopher, he was first known for writing a doctoral thesis that showed how with an exact language we can map reality (Wittgenstein, 1923). Then he spent the rest of his life trying to convince us that he was wrong—that there is more to human language and interaction than can be written down, and that language is action. Instead of focusing on mirror images of reality we are advised to think of the language games people play—how we are able to participate in human activities because we have learned to act according to the unwritten rules of that activity.

For example, if Pelle, a designer, points at the cardboard box with the sign "desktop laser printer," and says to Jon, a typographer, "Could you take a look at the proof coming out from the desktop laser printer," Jon does not answer "There is no desktop laser printer! You are pointing at an empty cardboard box, stupid." Rather he would go to the cardboard box, pick up a blank paper from a paper stack beside the box, turn toward Pelle, look at the paper, and say "Well, here we have a problem. There is too much text for a 48 point three column headline and that picture of the president. I think we will have to crop the picture, or the headline has to be rewritten."

According to Pelle, and the other participants, Jon makes a proper move in the design *language game* he is participating in. On the other hand, if Jon had maintained that there is only a cardboard box with a sign on it saying "desktop laser printer," he would have made an incorrect move in this specific design language game. Despite the fact that he would be right, he would not have understood how to play according to the rules.

The reason that Jon, Pelle, and the other participants can use the mock-up in a proper way is because this design language game has a *family resemblance* with other language games they know how to play. The language game in which the cardboard box is used has a family resemblance with the use of a traditional proof machine in the professional typographical language game which Jon knew very well, as well as with technical discussions Pelle had participated in as part of his profession. Furthermore, they both know how to play this design language game using the mock-up, because it resembles other *games* they have played before.

However, cardboard boxes do not become laser printers by themselves. In fact, one of the hardest challenges for the designer seems to be to *create a design language game that makes sense to all participants*; the designer in the role of play-maker. In this role the designer sets the stage by finding and supporting ways for useful cooperation between professional designers and "designing users." Future workshops and metaphorical design (Chapter 8), as well as organizational design games (Chapter 12), are examples of ways to set the stage for such shared design language games. Mock-ups and prototypes (Chapter 10) may be useful "properties" in these games. Hence, mock-ups are only effective in the design language games that make sense to the participants. In these games mock-ups play an important role as something to which one may refer in discussions of the design; as a *reminder* pointing back to experience from using the mock-up. Thus, instead of having to produce rational arguments in support of a certain point of view concerning a breakdown in the use of a mock-up, it is possible to repeat the sequence of operations leading to the breakdown. Then both that situation and the steps producing it may be evaluated, alternatives tried out, and, if necessary, participants may try to give rational arguments in favor of their point of view.

In summary, mock-ups become useful when they make sense to the participants in a specific design language game, not because they mirror "real things," but because of the interaction and reflection they support (see Ehn, 1989).

A new role for the designer is to set the stage and make it possible for designers and users to develop and use a common situated design language game. This has to be a language game that has a family resemblance with the ordinary language games of both the users and the designers; a language game which is socially constructed by the participants.

Hands-on Experiences and Ready-to-hand Use

There are, however, more to mock-ups and the language games in which they are used than just language. As opposed to linguistic artifacts, such as flowcharts and system description documents, mock-ups make it possible for the user to get *hands-on experience*. This is illustrated in the picture below. What you see is the first mock-up we ever made in a design language game.

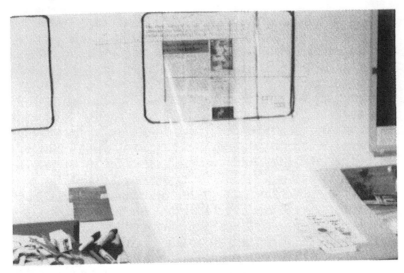

First mock-up of page make up work station.

In the above picture there is a high resolution graphic display, a control display, a tablet with a tablet menu and a mouse. Functionality of the system is simulated by making successive "drawings" of the screen. As shown in the next picture these drawings are "stored" on the wall, and "retrieved," "updated," and "changed" as the design game is played.

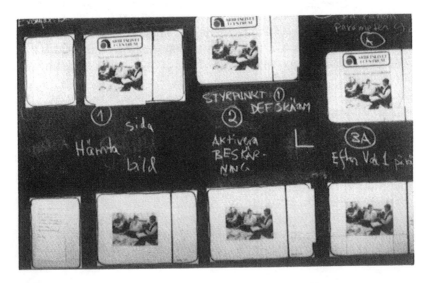

The wall as "store" for "interactive" display images.

This mock-up was the creative result of a major breakdown in the UTOPIA project, a breakdown that made us develop some new design artifacts and to shift perspective "from system descriptions to scripts for action" in participatory design (see Chapter 12). As designers we had been producing an endless number of detailed and methodologically correct system descriptions. There was just one problem. The users could not understand our system descriptions. The descriptions did not remind the users of familiar work situations. There was no meaningful role for them to play in the use of these design artifacts. The experience of using these descriptions did not relate to their work experiences. The mock-up above changed the rules of the game; it made it possible for the users to actively participate in the design process.

For example, Jon simply sat down by the mock-up and pretended that he was doing page make up work. He used the mouse, the tablet, displays and menus to crop a picture, move a headline, change a font, etc. This was done in a way that had a family resemblance to his traditional way of working. He understood the mock-up as he understood his traditional tools.

We take as an important starting point in design the idea that 'in the beginning all you can understand is what you already have understood.' In stating this design paradox we have been inspired by Martin Heidegger and existential phenomenology (Heidegger, 1962, and especially Winograd & Flores, 1986, and Dreyfus & Dreyfus, 1986). The point is that the mock-up did not create a breakdown in Jon's understanding. It was not present as an object in itself, but *zuhanden* (ready-to-hand) for him in his activity. Jon was primarily involved in page make up work, not in detached reflections over this activity. He was not reading or talking about a future system, but experiencing it as a *Zeug* (dress, tool, artifact) for page make up— he was literally well-equipped, rather than overloaded with equipment.

However, the mock-up is, obviously, not the same as his traditional typographical tools; hence, breakdowns in his readiness-to-hand use of the mock-up occurred. Typographic tools such as the knife became computer equipment such as the mouse and display; the mouse became a match box, the display a sheet of paper. When the spell of unhampered involvement is broken, the mock-up becomes *vorhanden* (present-at-hand) as a collection of things or objects. This is not an entirely sad story. After all, if the artifacts we use were always ready-to-hand for us, how could we then find new ways of using them? When things do not work, we shift to detached reflections of them. In the situation noted this meant reflections such as: "Is a mouse/display replacement of the typographical knife really a good design choice?" "Is the problem rather that the properties of the knife are too restricted, and that we with computer support can add some new useful properties like 'undo cut' and 'resize'?" These kinds of questions were part of the inter-action between the typographer using the mock-up and a designer sitting by his side. They certainly led to new design ideas, as well as to changes of the mock-up. For example, the second version of the mock-up provided possibilities for a wider range of hands-on activities and more elaborate design language games. There were more elaborated interaction devices to try out and a more dynamic interaction with the mock-up. Aspects of the work environment and of work cooperation could be tried out.

In summary, hands-on experience is not a substitute for detached reflection. However, in participatory design it is necessary and more fundamental to support the users' ready-to-hand use of their future artifacts. Hence, an important aspect of a mock-up is its usefulness for involved activity where the users' awareness is focused on doing the task, rather than on analyzing objects and relations. Detached reflections on alternatives become part of the process when the fluent use of the typographical design tools—their readiness-to-hand—breaks down. These reflections are then grounded in a practical experience, an experience shared by users and designers in a design-by-doing language game.

Beyond the Cardboard Computer

A "second generation" UTOPIA mock-up with a back screen slide projector.

The idea of "hands on the future" as opposed to "eyes on a system description" was our main focus in the previous section. We have discussed this in terms of mock-ups built from the most simple materials, such as cardboard and paper, but there are obviously other possibilities for getting hands on the future, most notably computer-based prototypes. Such prototypes are discussed in Chapter 10 by Susanne Bødker and Kaj Grønbæk. In this section we discuss some of the possibilities in the borderland between the "cardboard computers" and the "real" prototypes.

Second Generation Mock-ups

As a first step let us consider some possibilities that are more complex, although still not computer based. Depending on the availability and expertise in the design group, such possibilities may include overhead and slide projectors, tape recorders, and video. These artifacts are familiar in the sense that people easily distinguish malfunction in the artifact from malfunction of the design. At the same time they provide some useful functionality beyond that which is achieved with cardboard, and they make possible a "look and feel" that is more like the future product.

Designers and potential future users envisioning the future of page make up playing with the UTOPIA mock-up.

In our second mock-up of the text- and image-processing work-station in the UTOPIA project we used a slide projector and a screen for back screen projection. The first impression of this mock-up was much closer to the imagined final product: The display inter-action was simulated by the use of slide shows and the input devices had a "real" touch. This quality proved especially valuable when judged by people who only tried out the mock-up for very short periods of time. It was fairly easy for them to envision the future artifact by using the mock-up. This last point is important, and in many cases this alone may justify the use of slide projectors or video in a mock-up. The trade-off is that such mock-ups require more expertise and more resources, both in time and in money, and they are more difficult to change.

A sample of different key pad and "mice-like" input device mock-ups produced in the UTOPIA project in an attempt not to get stuck in the emerging standard interface.

Simple Mock-ups: Advantages and Disadvantages

Before we turn our attention to the use of computers in mock-ups, we will briefly outline some advantages and shortcomings of sim-pler materials to get a better understanding of what might be gained or lost from the use of computers.

   Until now we have looked at how to design *without* computers, not because we think that people should avoid computers in general, but because there are good reasons to think twice before using them, as well as good reasons to proceed "even" if computer support is not available. First, the mock-ups discussed so far are built with inexpensive materials. To buy expensive hardware and build advanced software early in a project may, in most situations, be directly counterproductive, especially given the possibilities of mock-ups. In other situations, however, the investments in hardware and software may not be a problem—PCs may already be massively used in the organization. Still, the use of mock-ups may pay off, because it can help generate new visions and new options for use.

Second, the characteristics of these simple tools and materials are familiar to everybody in our culture. With this type of mock-up nothing mysterious happens inside a "black box." If a picture taped to the blackboard drops to the floor everybody knows that this was due to difficulties of taping on a dusty blackboard, and not part of the design. There is no confusion between the simulation and the "real thing."

Third, such mock-ups lend themselves to collaborative modifications. The possible "operations" on the material using pens and scissors, for example, are well known to all, and with simple paper-and-cardboard mock-ups people often make modifications jointly or by taking quick turns. The physical changes are visible, and, with proper display, visible to all the participants.

However, as with any tool or technique, simple mock-ups have their limitations. Changes to a mock-up may be very time-consuming. If, for example, a different way of presenting menus is chosen, changes may have to be done to dozens of drawings, or a whole new set of slides may have to be made.

While it allows a design group to experiment without the limitations of current technology, this freedom is only a partial blessing. In the end, good design results from exploiting the technological possibilities and limitations creatively, not from ignoring them. Thus, as paradoxical as it may sound, the demand for computer knowledge in a design group using mock-ups is very high.

The simple mock-ups lack functionality: They represent physical clues with which one may create the illusion of using a future computer based artifact, but the users do have to use their imaginations along with the mock-up.

Computers in Mock-ups: Overcoming the Disadvantages

Now let's enter the borderland between "cardboard computers" and "fully computerized prototypes." In this borderland, distinctions between the two are fuzzy. In fact, we do not see the main difference between a mock-up and a prototype as being a question of whether computers are used or not. With mock-ups—computer-supported or not—the focus is on support for overall envisionment. In a powerful analogy to film production, this kind of envisionment has been called *storyboard prototyping* (Andriole, 1989). Marty Kline, the artist who drew the storyboard for the movie *Who framed Roger Rabbit?*, makes the analogy clear: "Storyboarding is a way to look at the film without spending a lot of money...It's not the ultimate film, but it represents a first chance to look at it" (Braa & Ruvik, 1989).

Moving from mock-ups and storyboard prototypes to real prototypes, the possibilities to demonstrate real computer-based functionality come into focus. Computerized prototypes differ from the use of computers in mock-ups in two important ways. We often use computers in mock-ups for purposes other than those intended for the future computer system. In the mock-up we are typically interested in using computers for envisioning the system, not to provide the real functionality of the system. Also, computers have no privileged position in relation to other materials such as cardboard and paper. They are all used on the basis of how well they contribute to creating the illusion of using the future system. In our investigations of the borderland we now consider if and how we may use computers to overcome the disadvantages described above, and if we may do so without sacrificing the advantages. We look first at the use of computers as a way to improve efficiency in building and changing mock-ups. Then we look at ways to explore technological limitations by means of computers. Finally we discuss the question of getting more functionality.

Effective Tools

The next pictures are from a recent project in which we are developing and using a computer-based hypermedia design environment that we call DesignSupport. To test and develop DesignSupport we have used it for design of a budget system. The budget system was intended to support our own research group in discussions on how to spend our funds. Most of these functions were not covered by standard budgeting and accounting systems, they were supported only by manual procedures using pen and paper, together with e-mail. The pictures are from an early mock-up/storyboard prototype of the budget system and show a scanned version of a handwritten economic overview, some links added to the drawing, and a "computer redrawn" screen image based on the handwritten overview.

A scanned version of a handwritten economic overview

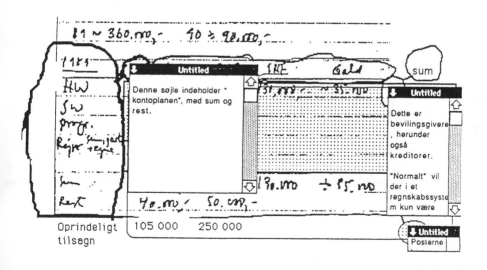

... *some links added to the drawing*

13/03/1989 Bevillinger er de penge vi har at gøre godt med i indeværende år og de følgende, så langt som vi nu har tilsagn.		13/03/1989 Til rådighed viser hvor meget vi har tilbage i dette samt de efterfølgende år tilsammen.			
Disse årstal fungerer som knapper, hvormed man kan gå til tilsvarende oversigter for de enkelte år.	e bevillinger med tilsagn/aftaler, f.				

	g	Forbrug	Disponeret	Til Rådig	Råd.h.-ønske
		nu	nu+ff	=B-F-D	
SoftWare					
Programmør					
Rejser&Sem					

... and a "computer redrawn" screen image based on the handwritten overview. All the material exists in the same "hypermedia."

After scanning the economic overview we used DesignSupport to create linked screen images. The computer-based drawing allowed us to take advantage of similarities between the pictures. For example, copying shared icons between screen images was easy, and repeated changes to an image did not reduce its quality as occurs when using paper. Furthermore, several screen images could share parts. Modifications made to a part on one screen image was automatically "cascaded" to the other images.

Finally, it is worth mentioning that intermediate designs were easily saved. This encouraged the exploration of alternative designs for screen images, since backtracking to earlier versions was almost instantaneous if the current line of design proved to be unsatisfactory.

The examples of efficiency gains discussed here cover only a small sample of the potential of existing computer-based tools. Depending on the system being designed, tools such as a presentation manager or a word processor may help improve the efficiency in making and changing a mock-up. One should remember, however, that achieving this efficiency gain usually requires that people in the design group be skilled users of the computer-based tools, otherwise the tools may get in the way of the job to be done, changing the focus from the mock-up to the limitations of the tools.

Creating Suitable Real Limitations

As noted, the demands on computer knowledge in a design group working with simple mock-ups are very high. When a cardboard box is used instead of a desktop laser printer, someone in the group must know about such printers in order to get the game going using the box as a laser printer, not as a box. In cases where this knowledge needs to be developed, computers may be used to investigate specific technological possibilities and limitations, as illustrated by the following example.

Investigations of resolution and response times using a real computer.

During the period in the UTOPIA project when we used simple mock-ups we acquired a few real computer workstations with 15" bit-mapped screens. The idea was to build one or more prototypes of the emerging design. But although the hardware was powerful, the software was poor, and the prototyping could never keep up with the mock-up work. However, it was useful to be able to experiment with a 15" screen as one of our mock-up components, especially because the existing knowledge in the design group on bit-mapped screens was not comprehensive. We began to look at questions such as: "how could a newspaper page be represented with the available resolution and screen size?", "how about a spread, that is, two pages?", "how many pixels were needed to make a font readable on the screen?", and "how about using shaded boxes to represent words in small fonts?" Such questions could not easily be dealt with using the paper images and slides of our first and second generation mock-ups, but they could be investigated quite easily using the workstations with their graphic screens.

Having learned about the possibilities and limitations of our 15" bit-mapped screens, we returned to our simple paper-based mock-up to explore the possibilities of different screen sizes, such as 15", 19", and 24". We cut holes of appropriate sizes in large pieces of cardboard and placed them on the wall in front of our pictures, menus etc.

... and expanding the screen size with a mock-up.

More Functionality

As the third and last of the disadvantages of mock-ups we want to address by means of computers, we look at the question of functionality. This question takes us very close to the borders of "real prototypes."

Consider once more the use of DesignSupport for the creation of a mock-up of the budgeting system discussed previously. The screen images could have been printed out and used in a paper-and-cardboard mock-up like any other picture. What we did, however, was to use the computer to show a sequence of images as a slide projector does.

The next step we took was to use the "button-capabilities" of HyperCard to make it possible from every screen image to dynamically select the next image, in a way simulating how this could be done in the final system.

Using the button capabilities together with text fields in constructing the screen images made it possible to simulate a number of dynamic changes: text-entry, selection, etc. Dynamic response from the budget mock-up, such as showing the money available for inviting guests for the rest of the year, together with the estimated cost of planned and "considered" visits, was handled by a human operator simulating parts of the system. Still other kinds of response, such as sorting a list of possible conferences to attend, were not handled dynamically but rather by showing a list prepared in advance, as we did with the paper-and-cardboard mock-up.

As the example shows, there are different ways to simulate functionality, and there is also the question of what functionality to simulate in the first place. There are no simple answers, but the yardstick to apply is how the different aspects contribute to the creation of the use situation envisionment, how useful it makes the mock-up in the particular design language game. Obviously there is a tendency to implement those aspects which fit the computer best, but the tendency to implement "computer-based functionality," as opposed to different kinds of simulated functionality, is quite strong, too. In the budget system case discussed above, our programmer discovered a clever way to program Xerox-like scroll boxes in our hypermedia system. Such boxes vary in size to indicate how much of a document is shown in a window. Viewed in isolation, this approach was superior to the existing Mac-like scroll boxes of fixed size. But at the time the question of the detailed workings of the scroll bars was unimportant in relation to the creation of a suitable use situation envisionment.

In summary, computers may be used to overcome severe shortcomings in the use of simple mock-up materials. But as we shall see the costs may be high; for example, in terms of reduced possibilities for user participation. However, there is often no need to "go all the way": the best and most cost-effective envisionment may well be obtained by a mock-up from the borderland between cardboard and computers, as illustrated by the following example.

Mixing for a Better Envisionment

Our first prototype of the design environment DesignSupport was built in HyperCard and ran on Mac IIs with 21" screens. The prototype was built over a period of a few months, ending up with an almost fully functional prototype. HyperCard, however, only allowed one 9" window to be open at a time, a restriction that turned out to reduce the usefulness of the prototype dramatically. The solution to this problem was straightforward, but we were so fascinated by the prototype that it took some time to find it: To be able to show several windows in varying sizes at the same time, we simply placed printed copies where we wanted them on the big 21" screens. This worked so well that Morten immediately began to "click" on them; forgetting that it was a mock-up. He was getting an involved experience of the future use even if the functionality was missing.

From weak prototype to strong mock-up by adding paper windows to the prototype.

This example illustrates an important difference between implementing a final system or fully functional prototype on the one hand and building a mock-up or a storyboard prototype on the other—a difference that seems to be forgotten easily once designers skilled in programming bury themselves in the computer. The point is that any design environment, computer-based or not, has limitations that at times place severe restrictions on the artifact being constructed. In the implementation of a new computer system the handling of this "tension" is a primary part of the competence of the professional designers. However, when constructing a mock-up, it is not necessary to restrain oneself to the possibilities of the computer-based aspect of the environment, unless the intention is to explore exactly these possibilities; for example, with the intention of implementing the final product using that environment.

We later implemented DesignSupport in an environment with multiple and re-sizable windows. That was a major improvement of the system as such, but our computer "blindness" for a long time prevented us from having these properties in our mock-ups and early prototypes.

Computers in Mock-ups: Losing the Advantages?

In the beginning of this chapter we suggested that the point in using non-computer-based mock-ups was that they are cheap, understandable, and allow for hands-on experience and pleasurable engagement. Certainly, computer-based prototypes encourage hands-on experience, and in many organizations hardware and software for prototyping already exists, so resources may not be the bottle-neck. Whether it is more fun to sit by a computer or to build with cardboard, we can only guess. The remaining advantage primarily concerns the understandability of the non-computer mock-up tools and materials: How does the computer fare with this?

Unfamiliar Tools and Processes

Consider the following situation, in which computer scientists from two geographically separate groups got together to work on the design of a "shared material" supporting joint work between their two settings. They decided to use two LISP machines on a network to quickly build a computer-based storyboard mock-up. Two of the computer scientists were LISP experts; the others were less familiar with the LISP environment. Since building and modifying the mock-up was a major and integrated part of trying it out in this first session, it was the two LISP experts who operated the machines. Thus, to the rest of the group the interface to the emerging design consisted of the two LISP experts. Involvement consisted mainly of discussing ideas and their possible embodiment in the LISP machines. Several of the actions carried out by the two operators involved programming for two or three minutes. During those periods the rest of the group was inactive.

The first thing to note is that the tools and materials used in this session were not familiar to all participants. Most of them did not know what could be mocked up, and certainly they did not know how to do it. In other words they had only vague ideas about the possible moves in this design game, and they could perform just a few moves themselves. Secondly, most of the "construction work" left no visible clues; thus, the status of the mock-up was not clear to most of the participants. The result was that after a short while only the two LISP experts operating the machines were able make constructive moves. The rest of the group had nowhere to place their hands.

This could also be viewed as an example of a badly planned process. The main point in design-by-doing using mock-ups is for everyone to get hands-on experience, trying something new. This acting in the future does not happen by itself. Especially with mock-ups built using unfamiliar tools and materials, the simulated future use situation has to be carefully planned and enacted.

What's the Purpose?

As our last issue we consider the expectations of people working with a mock-up, and what the purposes of doing it are. With cardboard mock-ups it's simple: the purpose is design, and the mock-ups are used to evaluate a design, to get ideas for modifications or maybe even radical new designs, and to have a medium for collaborative changes. If experiments with computer-based mock-ups are set up in the same way and their purpose made clear, it can be equally simple. But the functional possibilities may be seductive, especially when we approach the borders of functional prototypes. Often it is possible to build a computer-based mock-up/prototype which has the look and feel of "90% of the real system," and then the use of this mock-up is interpreted by the users, or maybe even set up by the designers, in a way that presupposes that it is "90% as *useful* as the final, real system." However, this is rarely the case. When this is realized by the users their interest in using the mock-up/prototype may easily drop or disappear completely.

As mentioned earlier, successful evaluation of a mock-up requires careful planning and acting, but in addition it requires commitment from the users, resources dedicated to the purpose of evaluating the mock-up. Almost any deviation from the final system in a mock-up requires some active work on behalf of the involved (future) users. If the users are not prepared to pay this price, then using the mock-up will fail.

Mock-ups: Prototype, "The Real Thing" or Both or ?

One of the reasons for the effectiveness of cardboard mock-ups is that nobody confuses them with the product, the future computer system; everybody knows that they do not have the functionality of a computer system. With computers in mock-ups it's different, especially when we use computers to get more functionality.

In these situations it may be difficult not to mix the appearance of the computer in a mock-up and in the imagined future product. The closer the two "roles" get, and the less familiar the computer is, the more careful one has to be in avoiding attributing the wrong aspects of the mock-up-computer to the future-product-computer.

Major Players and the Rules of the Game

Typographers, journalists, and designers in a game of soccer.

The picture above ends our story at the same place were it started—with the people missing in the first picture; showing the relationship between typographers and journalists. Here they are, in a game of soccer, and there are even some professional designers participating.

This game took place on a nice May afternoon in 1984. It was one of the activities that formed a workshop that was part of the "systems delivery" from the UTOPIA project. Typographers, journalists, trade union and management representatives were invited to actively participate in a three-day workshop on the design proposal from the project group. Of course, mock-ups and prototypes from the design work were tried out in hands-on sessions, but the "system requirement specification" certainly implied and included more than that. Not only the artifacts to be used were at stake; other aspects related to quality of work and product were also part of the proposal, especially questions of how work should be organized using these new tools, and what training and education the different groups should have.

On the soccer field typographers, journalists, and designers had no problem cooperating in mixed teams. This game was even more fun than playing with mock-ups. The negotiation game concerning the proposed changes of work roles and work practice was an entirely different story.

In fact, we had developed a useful work organization design kit to be used by the participants in this kind of negotiation situations (see Chapter 12), but that did not change the hard facts of reality: Some players have more power than others, and some are more vulnerable than others. For all the fun there is in design as action and in the use of mock-ups, implementation may be an entirely different game in which management prerogatives define the rules, and organizational conflicts between typographers and journalists limit forceful countermoves. In this game, often referred to as class struggle and organizational conflict, there is a temptation for the designers to think of themselves as observers just watching the game. Nothing could be more wrong in design as action, except perhaps the designers appointing themselves as referees of the game: the gods that make the other players obey the given rules.

As discussed in Chapter 7 in design as action, the rules are at stake. This is particularly true where the use of mock-ups is a way of experiencing the future. This is serious business concerning major changes of the participants' working lives. In using inexpensive mock-up tools and in establishing the pleasurable engagement of hands-on experience, the designers have to find their own role in the design game. The roles of observer and referee are not available. What defines the professional designer is the competence to find a proper role in a specific design game and to expand the space for users to participate in design as action.

References

Andriole, S. (1989). *Storyboard prototyping—A new approach to user requirement analysis.* Wellesley, Massachusetts: QED Information Sciences.

Braa, K. & Ruvik, E. (1989). *Edb-stötte til kreativt samarbeid i filmproduktion* [Computer-Support for Creative Cooperation in

Film Production]. Oslo: University of Oslo, Department of Informatics.

Dreyfus, H. L. & Dreyfus, S. D. (1986). *Mind over machine—The power of human intuition and expertise in the era of the computer*. Glasgow: Basil Blackwell.

Ehn, P. (1989). *Work-oriented design of computer artifacts*. Falköping, Sweden: Arbetslivscentrum and Hillsdale, NJ: Lawrence Erlbaum Associates.

Heidegger, M. (1962). *Being and time*. New York: Harper & Row.

Kyng, M. (1988). Designing for a dollar a day. *Office: Technology and People*, 4: 157-170.

Winograd, T. & Flores, F. (1986). *Understanding computers and cognition—A new foundation for design*. Norwood, NJ: Ablex.

Wittgenstein, L. (1923). *Tractatus logico-philosophicus*. London: Kegan Paul.

Wittgenstein, L. (1953). *Philosophical investigations*. Oxford: Basil Blackwell.

10

Design in Action: From Prototyping by Demonstration to Cooperative Prototyping

Susanne Bødker and Kaj Grønbæk

> ... *the development of any computer-based system will have to proceed in a cycle from design to experience and back again. It is impossible to anticipate all of the relevant breakdowns and their domains. They emerge gradually in practice.*
>
> Winograd and Flores, 1986, p. 171

Some time ago we worked with a group of dental assistants, designing a prototype case record system to explore the possibility of using computer support in public dental clinics. The application was not intended to be used directly in the treatment sessions, but to help administer the patients' visits. The dental assistants would be the primary users of the application, and they took part in a series of prototyping sessions to specify how they would like to use computers in their work. They knew that some kind of computer application would soon be introduced at their workplace, and it was important for them to be able to influence the choice of system. The following scene is taken from one of the prototyping sessions:

At one point two dental assistants were sitting together in front of the screen with the designer standing just behind them. They switched between screens containing pictures representing patients' teeth. Suddenly one of them said: "The lower jaw is turned upside down—that's quite confusing given how we number the individual teeth." The designer had turned the upper and lower jaws the same way, with the front teeth pointing upwards. The dental assistants did not like this. One of the dental assistants explained: "I want to think of the pictures of the teeth as I'm looking into the mouth of the patient when I look at the screen." The designer took the mouse and used a few menu operations to flip the picture. This gave all patients in the database the new, improved tooth representation. The modification activity lasted only a few minutes, without anyone touching the keyboard. The dental assistants were amazed by the ease of changing the prototype. They were satisfied with the change and one of them took the mouse again and continued using the prototype.

System Design and User Involvement

We saw that by actually using the prototype, the dental assistants were able to make changes that may not have otherwise surfaced until the users got the final system in their hands. The designer, of course, thought that he had turned the picture right. He had not realized that there was a "mapping" between the drawing and the position of the jaws in the mouth.

In this and similar examples, we find strong arguments for a more direct and active involvement of users in the design of computer systems; to find out how the computer application functions in the use situation the users must somehow be able to experience it. We call this *envisionment*. To experience is not to read a description of the computer application, nor is it to watch a demonstration. We have found prototyping to be very useful in uncovering unarticulated aspects of users' work and in having them contribute to the design of improved tools. In envisionment, breakdowns may lead to a change in the prototype, and eventually to a change in the future computer application. What we find useful in prototyping, relative to the use of mock-ups as described in Chapter 9, is that a prototype better shows dynamic aspects of the future application.

In this chapter we will discuss how to get started with prototyping that involves users actively and creatively. We will give examples illustrating how to obtain a close coupling between design activities and experimental evaluation of prototypes in work-like situations. Before going into the examples we give a brief overview of current prototyping approaches and point out how the approaches we propose are different.

Different Approaches to Prototyping

A rich variety of approaches to prototyping has been proposed in recent years. They all provide possibilities for users to gain hands-on experience before the final application is built; and yet, the way they are used today, they are not often applied this way. Grønbæk (1989b) gives a critique of three categories of prototyping: *prototype becomes the system*; *executable specification*; and *exploratory* approaches.

The *prototype becomes the system approaches* dominate current system development practice and literature. They supplement a traditional requirements specification with a prototype of user interface aspects before the implementation begins, primarily for the users to adjust details of the system. From empirical studies (Grønbæk, 1989a) we see that such prototypes are used primarily to demonstrate features, and not to let users try them out actively.

The main purpose of *executable specification approaches* is to obtain a full, formal specification of what (parts of) the future system should do. The specifications are made in a formal, executable specification language, meaning that a program, and thus a prototype, can be generated automatically from the specification. Although in theory, users and designers can evaluate the specification by evaluating this prototype, in practice this is hardly ever done. The specification languages that serve as basis for generating the prototypes are usually not suitable as a means for communication with users and empirically it has turned out that a full-scale formal specification of a system requires an effort similar to traditional programming of the system. Offsprings of these two approaches are seen in the so-called CASE-tools. Code generating CASE-tools have, however, not been used enough yet to assess their ability to support prototyping.

In *exploratory approaches* mock-ups, simulations and "throwaway" prototypes are developed employing various tools. The aim of the approaches is to make "quick-and-dirty" sketches of the computer application in order to clarify requirements for a new computer system. A number of cases describing user involvement in the evaluation of prototypes exists, but there are not many examples where users have been actively involved in the design and modification of prototypes and thus have creatively influenced the future system. However, the rapid development of simulated applications can facilitate early hands-on evaluation. The prototypes in these examples serve mainly as substitute specifications for the application and to propagate ideas into detailed design activities (see Bødker, 1987).

Cooperative Prototyping to Stimulate User Participation and Creativity

The traditional prototyping approaches mainly take the perspective of the designers and software engineers, and pay little attention to user involvement in the design process. We will now introduce a slightly different approach that we call *cooperative prototyping*. The approach has its roots in the exploratory approaches previously described, but we will demonstrate that prototyping can be a cooperative activity between users and designers, rather than an activity of designers utilizing users' more or less articulated requirements.

The cooperative prototyping approach aims to establish a design process where both users and designers are participating actively and creatively, drawing on their different qualifications. To facilitate such a process, the designers must somehow let the users experience a fluent work-like situation with a future computer application; that is, users' current skills must be brought into contact with new technological possibilities. This can be done in a simulated future work situation as described in the beginning of this chapter, or, even better, in a real use situation. When breakdowns occur in the simulated use situation, users and designers can analyze the situation and discuss whether the breakdown occurred because of the need for training, a bad or incomplete design solution, or for some other reason. Breakdowns caused by bad or incomplete design solutions could be rapidly turned into improved designs, reestablishing the fluent work-like evaluation of the prototype.

To fully experience the prototype, the users need to be in control of its use for some period of time, and to try it out in a use-like setting. The roles of the professional designers include anticipation of the use situation and building/cleaning up the prototypes in between the prototyping sessions. Ideally, cooperative prototyping should be performed by a small group of designers and users with access to flexible computer-based tools for the rapid development and modification of prototypes.

Examples of Cooperative Prototyping

We will present two examples. One is a cooperative prototyping process in which users participate directly in changing the prototype, using direct manipulation design facilities. This process is primarily a laboratory experiment. The second example of a cooperative prototyping process takes place in a real organizational setting, but without the close cooperation in making changes to prototypes. In the first example, HyperCard on a Macintosh is used, and the focus is on exploring a new graphical *user interface* for a dentist case record system providing direct representation of teeth on the screen. In the second example, the 4th Generation System, ORACLE, is used, and the focus is on exploring *organizational aspects* of a new computer system in an office that manages registration of incoming and outgoing mail for a large trade school. The examples illustrate very different cooperative prototyping processes, one focusing on technical issues and one on organizational issues, both of which need attention in a system development project.

Prototyping of Graphic User Interfaces

This example, mentioned in the introduction, deals with the development of a prototype of a patient record system for municipal dentist clinics (for more details see Bødker & Grønbæk, 1989). The end-users are mainly dental assistants working in clinics in public schools in Denmark. The purpose of these clinics is to provide regular dental checkups and treatment for school children. These clinics, as well as the work of the individual dental assistants, vary a lot, according to the size of the school and the organization of work at the clinic. All of the dental assistants in these clinics, and thus in the prototyping sessions, were women, most of them with no prior experience using computers.

This prototyping process was carried out by a designer together with a number of dental assistants in an educational setting where the main purpose was to learn about computer technology—about system design in general and design for dental clinics in particular. The topic was to explore problems in and prospects for developing a decentralized patient record system, combining administrative information with treatment-oriented information. The application was not intended primarily for use in the treatment session, but enough medical information was kept to fulfil the demands for statistical information from the Ministry of Health, and to help administer the patient's next visit.

The Prototyping Process

In order to prepare the prototyping sessions we designed an initial prototype with some modification of a prototype case record system for general practitioners (Bajlum & Nielsen, 1988). Pictures of human teeth and a set of fictitious test data were incorporated into the prototype. Later on, we augmented it with report facilities provided by Reports.[1] A few illustrative reports were prepared. We had some foreknowledge of the application domain, and the preparation activities required only two to three days of design work in order to provide reasonably stable prototypes for the sessions.[2]

---

[1] Reports is a program for retrieving information from HyperCard stacks and formatting it as a report for paper printout. Laying out reports is done by direct manipulation in an editor that is separate from HyperCard.

[2] The main benefit we got from reusing the system for general practitioners was a piece of code that helped us generate a set of threaded cards to compose the case record for a single patient. The rest of the prototype was developed from scratch.

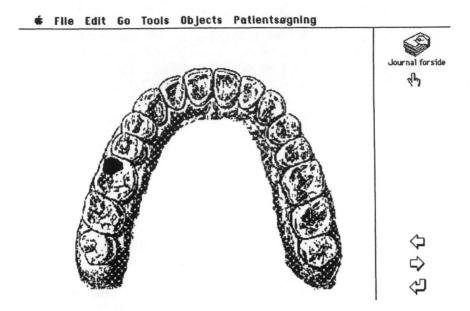

Figure 1. Direct representation of teeth in a prototype dental patient case record (Lower mouth part).

The dental assistants were on leave from their daily work to participate in the educational activities.[3] The prototyping process was introduced and discussed with the dental assistants, and the sessions started with a short demonstration of the prototype. Then the dental assistants worked with the prototype in groups of two to four to get a better understanding of it. The designer, who was one of us, had told the dental assistants that the prototype had been built in a flexible environment that allowed for changes to the design, encouraging the dental assistants to come up with suggestions for improvements. The designer tried to stay at a distance in order to let the dental assistants themselves explore the prototype and imagine that they were performing their daily tasks. There was no intervention by the designer in the process except in breakdown situations in which the dental assistants had problems or suggestions for improvements.

---

[3] According to agreements between unions and employers' organizations, shop stewards are guaranteed freedom to participate in trade union courses. Ordinary union members have no formal rights to participate in such courses, but in this case the union had decided to extend the offer to regular members. Thus the dental assistants' participation was paid for either by the union or by their employer.

Some groups worked quite enthusiastically with the prototype and came up with constructive suggestions, which were built into the prototype on the spot or after the sessions.

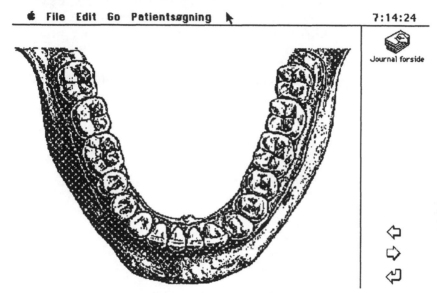

Figure 2. Direct representation of teeth in a prototype dental patient case record (Lower mouth part).

Examples of improvements that resulted from cooperation with the dental assistants in the prototyping session:

Change of teeth representation: The mouth pictures, which were used to indicate where treatment had been given, were initially turn-ed upside down, as described in the introduction, but HyperCard provided point-and-select tools to turn the pictures around.

Exploring alternative representations of tooth treatment: The dental assistants asked for direct marking of fillings on the mouth pictures as an alternative to our solution, which had each tooth as a button link to a separate card with a verbal treatment description. The visual aspects of this alternative representation could be explored immediately using the freehand painting tool of HyperCard, and thus be compared to the initial proposal.

Report lay-out design: The dental assistants participated in laying out simple reports from scratch and changing existing ones. This was possible because objects such as fields, text and graphics could be instantiated from available menus.

These modifications were done through direct manipulation; many such changes were made. The general experience was that the dental assistants became enthusiastic and creative when they discovered the potential for making these changes. We also made modifications that required modest programming. These included:

Copying buttons and modifying scripts: The dental assistants asked for functionality that was quite similar to functionality already available in an existing button. A few times we copied such a button and made minor modifications. When this programming lasted longer than a couple of minutes, the dental assistants lost their patience, because they could not follow what went on.

"Simple" query formulation and modification: When the dental assistants were participating in the design of reports they also had to participate in the formulation of queries. This soon became a little too hard. Not surprisingly they had to rely totally on the designers' suggestions regarding more complex queries constructed from ANDs and ORs.

A number of suggestions and ideas that came up during the sessions could *not* be integrated in the prototypes directly. The following are examples of this:

- The implementation of a menu to support searching for patients was made after the sessions, because it is rather time-consuming to program global menus. Moreover, the change could not be done via direct manipulation.

- In the sessions we only made reports with data from one type of card. One group realized that they wanted to print out a survey of the treatment of all the teeth of a patient. We had to tell them that while this ought to be done, it was quite a cumbersome process. Since programming reports was time consuming, the dental assistants didn't want it done during the sessions. For this reason many good suggestions for reports were never realized in the prototype.

Lessons Learned

Together with the dental assistants we came up with a number of conclusions about the prototype:

- The idea of cards with direct representation of teeth seemed promising.

- The idea of having each tooth on a separate card did not work for the dental assistants. They needed to be able to get a quick overview of all the treatment given to a patient. Such an overview could be provided with direct marking on the teeth combined with links to more detailed information. A similar combination of verbal and direct marking was needed to deal with treatments affecting several teeth. A major restructuring of the prototype seemed to be necessary to fulfil these demands.

Before the sessions we expected that the major challenge would be to keep the unreflected action of the users going in a prototyping process when, at the same time, we often had to stop and make changes to the artifact they were using. In our situation we could not go out into the real setting and have dental assistants use the prototypes there; our environment was not suitable for simulating a real dental clinic. It was the fact that the dental assistants were together that made it possible for them to "be in the situation" by having to demonstrate to each other how they did things "at home." It was possible for the dental assistants to step into the illusion that they could perform realistic work tasks with the prototype. The real challenge was to interrupt the process and modify the prototype, while they were criticizing some aspect of it, and then restart the evaluation process. Using a prototype that provided familiar cards and pictures of teeth made it possible for them to formulate very specific needs and requirements for a future computer application by using their own language and pointing at the prototype.

We expected that using HyperCard and Reports would often allow immediate design solutions to problems encountered with the prototype. As expected, problems arose when major HyperTalk programming was needed. Also, with the present tools, it was clear that cleanup of the prototype versions was needed. There was still quite a lot of work for the computer people to do in between meetings.

We see direct manipulation facilities as provided by HyperCard/-Reports as important for a cooperative prototyping process with in-session modifications. The direct representation of the data structure on the screen (the cards, fields, etc.) was valuable in this case. We can easily see situations where the card structure would be a limitation, but in this case the direct mapping of "good old-fashioned" cards on the screen made it relatively easy for the dental assistants to understand what the prototype could do for them and how the direct manipulation changes affected it. We experienced two sources of undesirable breakdowns in the design situation: reaching the limits of direct manipulation possibilities, and making changes that require major restructuring of the prototype.

These experiences show that it is possible for a group of workers, when given the chance, to come up with constructive and creative contributions to the design of their computer applications. The discussions quite easily got focused on the current prototype and rather technical issues, though. But regarding work-oriented issues the users in this case found out by experiencing a prototype that treatment-oriented work could be supported much better than with systems currently in their workplace.

Furthermore, the study illustrates that it is not necessary to aim at a prototype which simulates full functionality. It is possible for users to abstract from knowledge about these matters where necessary when provided with a prototype with some essential functionality. The initial prototype was stable enough to be evaluated in a work-like situation, and flexible enough to allow in-session modification. In the case just described, we made work-like evaluation and in-session changes of prototypes together with users in a laboratory setting. The following section describes a case where prototypes were tested in a real organizational setting.

Prototyping in an Organizational Setting

In this case, we explored hands-on prototyping utilizing a 4th generation system, ORACLE, taking prototyping out of the laboratory setting and into an actual work setting. It deals with a small group of workers in a large organization. We worked with the administration of a large Danish trade school, spread over a number of locations. This administration takes care of budgets and other financial issues, management of buildings and other facilities, including construction work, registration of students, salaries, and other staff administration, supplies, secretarial work, etc. Many of these functions are located partly centrally and partly locally. This administration carried out a large project with one of us as a consultant. The purpose of the project was to create an integrated office automation system, to provide more efficient administration of the school. The office administration system should be financed not by laying off employees, but by allowing for more efficient use of resources such as classrooms and heating.

The project was initiated by the school management. According to the local technology agreement,[4] it was managed by a technology committee with representatives from management and staff. It was the general idea of the project that the employees should, in project groups, take part in designing the computer applications that they are to use themselves. The school hired a number of consultants to work with the employees in the design work. The actual realization of the computer applications was carried out by a computer manufacturer on the basis of the specifications and prototypes created by the users and consultants. The case described here deals with one of these project groups, working with information storage and retrieval. The purpose of the group was to reorganize the files of the

---

[4] Technology agreements are agreements made between unions and employers to regulate the development and use of new technology within certain use domains. Refer to Mathiassen, Rolskov, & Vedel (1983) for details on laws and technology agreements.

school to be more efficient, eventually by means of a computer application.[5]

The file of all incoming and outgoing documents represents the history or memory of the organization. The information retrieval was rather cumbersome in the then-existing structure. The office where the file is located served the case workers in the administration, who acquired documents on specific issues from the file. The project group consisted of the women working in the filing office, representatives of the case workers who were the users of the file, and consultants competent with respect to both organizational issues and computers. There was a general agreement among the users that the way documents were filed and retrieved was not working well— each group of workers had problems with the system.

The Prototyping Process

The consultants began by interviewing the participants and observing the work in the filing office. Then the group gathered to start its part of the work. After initial discussions, three alternative ways of filing were proposed. Scenarios were made of how the future would look with each alternative. In addition, different ways of organizing the file and of doing the filing and retrieving were envisioned by the group. This was done through visits to trade shows and other workplaces in which filing technology had been applied. The filing office even borrowed a couple of filing systems from vendors and installed them in the office for a couple of weeks. This allowed the group to focus on what it wanted and what it didn't want from these different applications and the organization of work they required.

After this, an initial paper-based sequence of screen image simulations was discussed in the group. This mock-up illustrated one of the three alternatives: a paper-based file with lists of incoming and outgoing mail kept and distributed via the computer (see Chapter 9 for a discussion on mock-ups).

On the basis of these discussions, the consultants built a first prototype using ORACLE. Some questions to be considered were: What does it mean to have the lists of incoming and outgoing mail computerized? What information needs to be entered, and by whom? Which lists should be available to each caseworker? How should the lists be structured? Who "owns" the lists? At this stage there was a strong emphasis on the cooperation between the filing office and the caseworkers, on the different caseworkers varying

---

[5] For a more detailed discussion of the project and the specific case see Kristensen, Bollesen, & Sørensen (1986).

needs for information, and on the qualifications of the filing office workers.

As the versions of the prototype stabilized, they were also used in the real use setting—the women in the filing office used the prototype to create mailing lists and make these available to the caseworkers.[6]

Programming the prototype required a lot of experience because much of what the group wanted, concerning cursor movements and the like, could not be programmed directly in the available version of ORACLE, but had to be programmed in Pascal. The prototype, as it was, illustrated only a limited part of a future new application, a part that was running rather well. In the next step, when we wanted to expand this prototype, the database had to be completely restructured and the previous prototype, including test data, thrown away. We found that office management, in particular, was reluctant to take this step. Although the prototype was unable to handle the large data sets needed for daily work, management thought of the prototype as something running perfectly well in the setting. They tried to use it for work purposes as it was, and of course it failed. This issue is discussed in more detail later in a subsection on unrealistic expectations.

ORACLE allowed the group to run the prototype on the computer that was used for other purposes in the office. For some weeks the prototype was used daily, under supervision of some of the consultants. They helped each worker get started, they were there when things broke down, and they made observations of the use processes. The aspects that were illustrated had to do with how mailing lists worked. This prototype allowed experiments with a rather limited but essential part of a future filing system in the real setting. What could be experimented with was the communication between the caseworkers and the filing office, which involved a change in the traditional way of making mailing lists. There also came to be a strong focus on the qualifications of the filing office in relation to the work of the caseworkers: How much do the filing office workers have to know about the work that the caseworkers are doing in order to fill in the proper information in the mailing lists?

Lessons Learned

We learned from this process that using prototypes as well as existing systems as alternative suggestions for the future allows the par-

---

6 The workers denote various collections of task as "cases" and they label the person who takes the main responsibility to treat the case as case-workers or case- handlers. Thus we selected the term case-workers for talking about these employees in general.

ticipating users to formulate their suggestions better. It was not necessary to come to a consensus about the understanding of the problem as long as some solutions could be found that made everybody comfortable with the future use. By integrating the prototypes in the organizational setting it became possible to focus not only on individual use, but also on cooperation among people. And still, to avoid an overly technical focus, it was necessary to shift between techniques, and to bring in other kinds of "prototypes," such as the filing systems from other domains. The prototype's actual use required a robust prototype, running on equipment that could be made available in the use setting. Because the prototype ran on the same computers as other programs that the workers used, it was easier for them to get started, and easier to integrate it into the daily work tasks, than would have been the case with a prototype running on a separate computer. In this case the price paid was that it was difficult to actively involve the users in the actual construction of the prototype, because the programming effort was too large. The 4th generation system used was too limited in the facilities provided, the concepts used were too hard for the users to understand, and a structural change of the prototype was very difficult. Furthermore, in this case, a major restructuring of the prototype was prevented by the inflexibility of the 4th generation system.

It was important for all the participants to keep in mind the status of a prototype—the purpose it is intended to serve and the aspects of the future application it is illustrating in order to avoid giving rise to unrealistic expectations of the sort experienced by management in this case.

We have seen a case in which cooperation issues were important and in which prototypes running in the organization, as well as borrowed computer applications, were important elements in a cooperative design process.

How to Get Going with Cooperative Prototyping

We have given examples of cooperative prototyping based on the use of existing tools. To get a cooperative prototyping process going it is crucial for the working group to establish a common understanding of the aims of the process, the status of the intermediate products developed in the process, and the role of prototyping in the overall design process. Furthermore, some organizational problems must necessarily be handled to establish a basis for performing cooperative prototyping in a specific project. These problems cannot necessarily be handled within the project; some of them need to be handled before the project is established. In this section we summarize some of our experiences and suggest possible steps for system designers to take in order to get going with cooperative prototyping. We present these suggestions by pointing to a number of issues, or rather, tensions between issues that are worth considering.

Establishing Project Groups: Making a Workable Group versus Involving a Large User Group

There are a lot of issues involved in selecting a competent group of users to participate in design/prototyping activities. Many are not within the control of the designers, but are determined by power relations, technology agreements (in Scandinavia), and the like. System development literature (Harker, 1988; Pape & Thoresen, 1987; Grønbæk, Grudin, Bødker, & Bannon, 1990) and practice indicate that there are several ways to select participants. For example:

- middle managers who have an overview of the task domain;
- participants who constitute statistically representative samples;
- representatives elected by the users;
- employees with experience using computers;
- the most skilled workers among the future users;
- the most enthusiastic among the future users.

In addition the design can start as a pilot project in one department.

We are unable to point to one of these criteria as the most important, nor do we believe in setting up a single criterion. The appropriate choice in the actual project should be discussed carefully when establishing the group. However, we consider the first criterion mentioned particularly dangerous in relation to our view of design. Middle management, who are only involved in the tasks on an abstract level, cannot be expected to make relevant contributions through hands-on evaluation of prototypes, as they do not have the necessary familiarity with daily work processes. We say this even after encountering, in a previous project, the problem that workers at the shop floor had very little understanding of the planning and coordination of their work processes. In relation to establishing a working group of a reasonable size the second criterion is problematic too, because many computer systems have user groups which are quite big and diverse. Thus a statistically representative sample will result in too big a working group. What we find most important, though, is to establish a working group together with *competent user representatives*.

It is also important that the participating designers and users get to know each other quite well, because cooperative prototyping is based on the assumption that contributions from all participants are important. Cooperation is crucial to maintain the ongoing mutual learning process between users and system developers. Steps to be taken in establishing a project working group are discussed further in Chapter 7 and in Andersen, Kensing, Lundin, Mathiassen, Munk-Madsen, Rasbech, & Sørgaard (1990).

The product of the prototyping process is more than a computer prototype. Prototyping is a learning process, and much of the new understanding must be spread to workers and managers who are not participating directly in the prototyping process. One way is to use the different prototypes in a process in which all involved personnel are guided through an abbreviated version of the prototyping process (Bisgaard, Mogensen, Nørby, & Thomsen, 1989). In general, the prototypes are valuable means of educating future users, because education can start while the final computer application is being implemented. Similarly, the participants from a prototyping process may be able to act as teachers.

Depending on the size of the organization, a process such as the one described by Pape and Thoresen (1986) may be appropriate. They describe a process in which an intermediate prototype is built in one part of the organizational setting. The designers move on to a new part of the organization, bringing this prototype to be used as a starting point there. A new process with a new group of people is conducted, and the designers move on to yet another part of the organization, where, in this case, the final prototype is developed. As with all prototyping processes this requires mutual learning in each of the settings.

Setting up Prototyping Sessions: Designers as Conductors versus Users Being in Charge

To users, designing a new computer application is secondary, whereas for designers it is their primary work. This means that the designers must know how to set up the process, and must make sure that they all get something out of their meetings. This involves considering such questions as: What is the purpose of the session? How stable should the prototype be in advance? To what extent should in-session modifications be made? What setting should be chosen? How should the outcome be documented/evaluated? These issues are discussed in more detail elsewhere (Grønbæk, 1989a). A remark on the question of stability and allowing in-session modifications is relevant here. As we saw in the trade school case, cooperative prototyping does not require that a prototype be modified in the session, but, in our experience, using direct manipulation for modifications is a good way to engage the users. However, only some prototype modifications *can* be done in-session, and it is necessary that designers be aware of the constraints and know when to postpone a modification until after the session. In particular, much homework must be done by the designers to prepare for prototyping sessions. Thus, the designers still control much of the decision-making process.

In prototyping sessions, designers often like to demonstrate all the features they believe are wonderful. Our claim, however, is that demonstrations do not necessarily tell the users anything about how the prototype or the final application fulfils their needs. To fully experience the prototype, the users need to be in control of its use for some period of time—to try it out in a work-like setting. If the prototype is not sufficiently stable to let the users work on their own with it, the designer should be prepared to give first aid for breakdowns caused by the prototype. While designers are, initially, the ones who know how the process should be set up, it is important that the process be adjusted to the needs and wishes of the users.

Providing Prototypes: Showing Fantasy versus Being Limited by the Tools

We have given examples of how we used two quite different tools for prototyping, neither of which was ideal. Getting started, so that the users can experience using some prototype of the future system, is important. We find the experiences of potential problems and possible solutions gained from the early prototyping experiments quite valuable and thus worthwhile. The prototyping activities can always be supplemented with more flexible mock-ups and even traditional description to cover aspects of a future application that cannot easily be covered by the prototypes.

If unstable or poor prototypes are presented to the users there is, of course, a danger of missing the point. If, however, there is no realistic possibility for making better prototypes, the users' hands-on experiences with imagined parts of the future system are still valuable. In the trade school case we saw that existing but different computer applications can serve as sources of comparison to developed prototypes. Thus we could say that a poor example is better than no example, if the status of the prototype is made clear. We recommend that the shortcomings of a prototype be compensated for by a good explanation of the deficiencies.

Today a number of tools exists that can be utilized for cooperative prototyping (see Grønbæk, 1989b). Such tools are often available on PCs or graphic workstations, and they are not necessarily expensive. Smalltalk and various LISP tools can also be bought for PCs at reasonable prices. Some tools are useful for prototyping in certain application domains, although they are not application specific from the outset. For example, in the domain of patient case records, HyperCard, providing an interface based on a card metaphor, supports the prototyping of some aspects quite well, because the case records in that example did consist of cards in folders. To summarize, we find that it is possible to accomplish a lot with existing tools.

Finally, alternatives can stimulate the users' imaginations and thereby encourage the group to discuss different ways of organizing work. Exploring alternatives is not a waste of time, but a necessity to get fantasy into the development process and thereby to improve the users' work through better computer support. A way to stimulate users to propose alternatives is to illustrate the ease of modifying a prototype, as we did in the dental assistant example. Building a toolbox of elements to make exploration of alternatives easy would also seem to be a good investment in most cases; for example, general elements could support different interaction styles and devices.

Maintaining Communication: Describing Requirements versus Experiencing Work

Users are not there to annoy the designers or to spoil their "wonderful" design, but to guide them, because the users know the relevant work tasks. Users are not necessarily good designers of computer systems, but even awkward suggestions may be grounded in tacit knowledge related to aspects of the work that the designers do not understand fully. As discussed in Part I, designers will perhaps have to study the work of the users more carefully and discuss it with them further before suggestions can be understood or turned down. Keep in mind that the users are the key to the design of a useful system and the designers are the key to propagating the user demands into the technical design of the system. The delicate balance may well be to design an application that is both useful to the users and of high quality from a technical point of view.

The most fruitful communication of design ideas occurs when problems are explained within language games familiar to the users. Examples of this are seen in the case with the dental assistants. We used a direct representation, a picture of the teeth on the screen, instead of burdening them with explanations of a form or a record data structure. At the trade school we used concepts from the existing file and mailing lists. The principle used for the interaction with the system was similar to the idea of putting information on cards or sheets in a case record folder in the manual case records. In general we have found it useful to have users and designers experience the use of prototypes in work-like situations and to make use of language games that are familiar to the users.

Users Perception of the Process: Realistic Prototypes versus Unrealistic Expectations

To obtain a work-like evaluation in prototyping sessions, prototypes need to be realistic and stable enough to let the users be in charge of the evaluation. But it is also important from the beginning to create awareness of prototypes as rough drafts that should be thrown away if they are bad. Not every prototype the group comes up with will be a hit, and many times the group will choose a minimal solution when making modifications. The group may know all along that this solution is temporary, but it may still be worth pursuing.

However, prototypes direct the expectations of both designers and users in a way that creates a blindness toward other and maybe better ways of dealing with the issues being considered. The designers may expect that the prototype can be included in the final product. Alternatively, as at the trade school, users may find the prototype so stable that they think of it as the final product when it is not suitable for handling production data. A manager may, for example, demand that production data be entered into the database and treated by the prototype. These observations lead us to stress the need for setting the scene and creating awareness of the objectives of prototyping approaches.

Maintaining Focus: A Technical versus a Work-oriented Focus

Technical skills have traditionally been considered the important skills in the development of computer applications, so working groups have often consisted of only technically skilled designers and a representative for the "customer." However, as the first part of this book points out, the major problems are not purely technical, but rather grounded in the coupling between technical solutions and work, the organization of work, and political issues at the workplace. Thus, early prototyping experiments should act as catalysts for focusing on this coupling, as illustrated in the trade school example, in which prototypes were evaluated in the use domain and cooperation issues were in focus. To maintain this focus requires more than the purely technical skills needed to build prototypes. Domain knowledge is crucial, as is a certain level of knowledge of organizational issues in general. The question is whether we can realistically have it all. In our experience, the enthusiasm of being actively engaged in design may well be spoiled if the working group grows too big.

Prototyping naturally leads to a technical focus on the computer application, as was seen in the dental assistants case. There is a danger that the designer gets more focused on "hacking" than on discussing problems related to the use situation. Discussion in a prototyping session often focuses on the interaction with the computer more than on how work is organized around the computer. These matters are important (Bødker, 1990), but they are not all. To avoid a too technical focus it is important, as we saw in the trade school case, to shift back and forth to techniques that have different foci. And it is important to use experiments with the prototypes in work-like settings to get to an understanding of changes in work, such as in the qualifications and organization of work around the computer application. In our experience it is important to combine prototyping with other techniques, and not be afraid to step away from the prototype.

Getting Resources: Adding More Resources to Early Activities versus Lowering Development Costs

By highlighting the benefits of cooperative prototyping early in a development project we also point to the necessity for development projects to allocate more resources for these early activities. What might require the most resources, in contrast to traditional projects, is the participation of more users. To ensure real, active involvement, users have to be freed from part of their daily work to avoid requiring them to do double work. Yet it is also important that the users keep some involvement with everyday work tasks so that they do not become managers or professional user representatives.

A number of authors (among them Boehm, 1981) have pointed out that the most expensive mistakes or shortcomings in system development are those made in the early phases in which the focus is on analysis and design. At the same time, it is pointed out that the resources spent in these phases are traditionally the smallest portion of total expenses. We agree with these authors that there is an imbalance between importance and resources. We believe that cooperative prototyping can help anticipate some of the expensive "mistakes" and most likely reduce development costs (Lantz, 1986).

Design as an Ongoing Process

We have argued that cooperative prototyping approaches can attack a number of problems traditionally related to lack of user involvement in system development. Of course, they cannot solve all such problems; they will not resolve conflicts in organizations or indicate when prototyping has produced sufficient clarification to implement a system.

But cooperative prototyping can be applied to improve the quality of computer systems seen from the point of view of the users; that is, users can get a system that is more tailored to their needs and thus improve the quality of their work. However, organizations are not static, and technological innovations increase the number of different interaction styles that can be provided for essentially the same function. Neither designers nor users can gain full knowledge of the possibilities for providing computer support for an application domain; design is an ongoing process. Thus cooperative prototyping should not be viewed as an approach to produce the ultimate computer system for an application domain. It should rather be viewed as one of the initial steps in the ongoing development process in which a computer application is augmented and tailored in conjunction with changing needs. Actually, the possibility for users to create individual interaction styles and augment existing applications based on their own needs may not be that unusual in the future. Such aspects of computer systems are referred to as adaptability or tailorability; these concepts are discussed further in the next chapter.

Acknowledgments

Thanks to "our real user" John Kammersgaard and to many of the contributors of this book for useful comments.

References

Andersen, N. E., Kensing, F., Lundin, J., Mathiassen, L., Munk-Madsen, A., Rasbech, M., & Sørgaard, P. (1990). *Professional system development—Experience, ideas and action.* Englewood Cliffs, NJ: Prentice-Hall.

Bajlum, T. & Nielsen, T. (1988). *Edb-støtte til lægepraksis—et eksperiment med udvikling af en brugsmodel og en prototype* [Computer support for medical practice—an experiment with the development of a use model and a prototype]. Unpublished master's thesis, Aarhus University, Aarhus, Denmark.

Bisgaard, O., Mogensen, P., Nørby, M., & Thomsen, M. (1989). *Systemudvikling som lærevirksomhed, konflikter som basis for organisationel udvikling* [Systems development as a learning process, conflicts as the origin of organizational development]. DAIMI IR-88. Aarhus, Denmark: Aarhus University, Computer Science Department.

Boehm, B. W. (1981). *Software engineering economics.* Englewood Cliffs, NJ: Prentice-Hall.

Bødker, S. (1990). *Through the interface—A human activity approach to user interface design.* Hillsdale, NJ: Lawrence Erlbaum Associates.

Bødker, S. (1987). Prototyping revisited—design with users in a cooperative setting. In P. Järvinen (Ed.), *Report of the 10th IRIS Seminar* (pp. 71-92). Vaskievesi, Finland. Tampere, Finland: University of Tampere.

Bødker, S. & Grønbæk, K. (1989). Cooperative prototyping experiments—Users and designers envision a dentist case record system. In J. Bowers & S. Benford (Eds.), *Proceedings of the First European Conference on Computer-Supported Cooperative Work, EC-CSCW* (pp. 343-357). London.

Grønbæk, K. (1989a). Rapid prototyping with fourth generation systems—An empirical study. *Office: Technology and People, 5* (2), 105-125.

Grønbæk, K. (1989b). Extending the boundaries of prototyping—Towards cooperative prototyping. In S. Bødker (Ed.), *Proceedings of the 12th IRIS conference* (pp. 219-239). DAIMI PB-296-I. Aarhus, Denmark: Aarhus University, Computer Science Department.

Grønbæk, K., Grudin, J., Bødker, S., & Bannon, L. (1990). *Improving conditions for cooperative design—Shifting from product to process focus.* DAIMI PB-331. Aarhus, Denmark: Aarhus University, Computer Science Department.

Harker, S. (1988). The use of prototyping and simulation in the development of large-scale applications. *The Computer Journal, 31* (5), 420-425.

Kristensen, B. H., Bollesen, N., & Sørensen, O. L. (1986). *Retningslinier for valg af faglige strategier på kontorområdet—et case studie over Århus tekniske Skoles kontorautomatiseringsprojekt* [Guidelines for trade union strategies in the office area—A case study of the office automation project of the Aarhus School of Polytechnics]. Unpublished master's thesis, Computer Science Department, Aarhus University, Aarhus, Denmark.

Lantz, K. E. (1986). *The prototyping methodology.* Englewood Cliffs, NJ: Prentice-Hall.

Mathiassen, L., Rolskov, B., & Vedel, E. (1983). Regulating the use of edp by law and agreements. In U. Briefs, C. Ciborra, & L. Schneider (Eds.), *Systems design for, with and by users* (pp. 251-264). Amsterdam: North-Holland.

Pape, T. & Thoresen, K. (1987). Development of common systems by prototyping. In G. Bjerknes, P. Ehn, & M. Kyng (Eds.), *Computers and democracy – a Scandinavian challenge* (pp. 297-311). Aldershot, UK: Avebury.

Winograd, T. & Flores, C. F. (1986). *Understanding computers and cognition: A new foundation for design.* Norwood, NJ: Ablex.

11

There's No Place Like Home: Continuing Design in Use

Austin Henderson and Morten Kyng

> ... *effective design involves a co-evolution of artifacts with practice. Where artifacts can be designed by their users, this development goes on over the course of their use.*
> Suchman and Trigg, Chapter 4, p. 72

One of our primary goals in this book is to contribute to the creation of systems that better fit work situations, as well as the aims and intentions of the people using the systems. This implies a view of design as a process that is tightly coupled to use and that continues during the use of a system.

The preceding chapters in this part of the book have focused on the first of these aspects, showing how to incorporate use experience in the initial design process. This chapter will show what may be involved in continuing design in use, and how we, in the initial design process, may create systems are tailorable.

Case I: Tailoring?—Never Thought About It!

At Aarhus University, personnel in the payroll office were accustomed to helping people who had difficulty understanding their payslips. Because each person in the office handled only a few types of contracts, they understood the contracts thoroughly and could explain any peculiarities. However, due to the introduction of a new computer system, mandated for all state-owned Danish enterprises, the work was reorganized. As a result, each person at the payroll office handled employees according to their date of birth. This may have been the standard way of doing things at many public payroll offices, and thus an acceptable choice for the computer system, but at Aarhus University it resulted in a dramatic decline in the service provided to the employees; each person at the payroll office now handled so many different kinds of contracts that they no longer knew the details.

Although the change necessary to remedy the situation—to return to the "contract way" of working—seems conceptually simple in terms of the work being done, it was in fact infeasible. First of all, there was no support for "local variations" in the initial design of the system. Secondly, the process of maintaining the system didn't incorporate any notion of modification beyond the fixing of bugs.

Case II: Handling Modifications

As our second case, consider the Danish hospital that recently installed a new computer-based heating and ventilation control system built from a dozen standard components. The boilerman in charge of the daily operation of the system handled it quite easily. However, the personnel on duty outside normal working hours were not boilermen, and they had difficulty understanding the messages from the system and reacting appropriately. The problem was solved by having some of the "personnel on duty," together with the boilerman, redesign all messages from the system, rephrasing them in language they all understand. This second set of messages was then added to the system as the new default. As the personnel actually encounter the messages in their use of the system, those that still create problems are noted and usually modified.

In this case the system in question was prepared to accept messages in different languages, and this, in combination with a design process that deals with a number of modifications to the standard components, made the change quite feasible.

Case III: Preparing the System for Change by Its Users

As our last case we consider workers at three cooperating research facilities who are using the Xerox Common Lisp environment (XCL) for electronic paperwork, e. g. handling documents and electronic mail (cf. Sheil, 1983, for information on Lisp environments). These workers include members of the research staff, secretaries, and administrators, only some of whom can program in LISP. Although XCL is a very flexible environment for those who can program in LISP, it can be quite unfriendly to those who can't.

The researchers created Buttons to help (MacLean, Carter, Lövstrand, & Moran, 1990). A button is a screen object (actually a small window) that can be easily moved around, copied, and sent to others in documents and electronic mail. Each button contains an "action," an arbitrary piece of LISP code that is executed when the button is "pushed" by clicking on it with the mouse. Buttons are used to add to the behavior of the XCL environment.

Consider, for example, what happened when users noticed that considerable effort was being expended on remembering and typing addresses for people being sent electronic mail. A local supporter created a button, labelled with a (usually short version of) person's name, that simply typed in that person's address. Soon these "address buttons" were being copied and shared with others. Initially, users either got help in modifying buttons to capture the addresses of new people or learned how to do it themselves. The buttons were subsequently modified to make changing labels and addresses much easier. XCL was tailored by means of Buttons to meet the needs of electronic mail in ways not anticipated by its designers.

Ideal tailorable systems are those in which there are means for the users, or supporters near the users, to make them fit different work situations. The system we call Buttons is a very general tailoring mechanism which is, with varying degrees of ease, usable by workers to augment XCL behavior.

Professional system designers can match systems to users, because they have the skills to modify them. Creating systems where nonprofessional designers can do the same thing is not nearly as easy. So it is not surprising that program-literate researchers created the first address buttons. However, it should be noted that the nonprogrammers have a big advantage, in that they, as users, know the work the system is supporting, and are therefore in a much better position to determine when the system matches the work.

Why Do We Need to Continue Design in Use?

There are three main reasons why system behavior may need to change after its initial design is implemented. They are:

- As designers of technology we are usually confronted with the task of designing systems that will be used for long periods of time. And no matter how well they may have fitted the situation initially, circumstances of *the situation of use changes:* the needs change, the uses change, the users change, the organization changes. Therefore the computer systems may well have to change to match the changed circumstances. In the case of Buttons, an increase in the complexity and number of mail addresses changed the users' needs.

- The *complexity* of the world makes it difficult to anticipate all the issues that will eventually be of importance in the final situation. There are bound to be things we overlook, misunderstand, etc. The creator of the XCL electronic mail system, for example, did not provide a means for aliasing addresses (as is provided in other mail software); the address buttons were created to meet the need.

- When we are creating a product which will be purchased by many people, the need to design for many *different situations* of use is particularly clear, and in fact some of the techniques for allowing users to adjust systems in use have been motivated by market considerations; for example, producers of "plastic" software who want to satisfy as many users as possible with a single product.

Why Is It Difficult to Continue Design in Use?

Most of this book is about designing systems that fit the work of the users, and the difficulties involved in the quest of concepts and understanding that further such a task. When we want to support continued design in use, we face the same difficulties, magnified by the additional requirements that:

- the system include tools for doing this continued design;
- the continuing design be feasible for nonprofessional designers (the users);
- the "sub-system" for tailoring matches the users' work, just as required of the rest of the system.

But as with traditional systems, tailoring is more often based on concepts from the computer field than from the application domain.

First of all, the *need* for providing possibilities for tailoring has to be recognized in the development of a system. Otherwise such possibilities will be missing, or tailorable only in the implementation environment. In the development of "plastic" software this need to provide for tailoring is usually recognized. However, it seldom goes beyond the setting of software-switches (parameters), as discussed later. In custom development there is a tendency to try to provide "the right system," without regard for possible future changes. In this respect custom development may learn from plastic software design.

Recognizing the need, however, is only the beginning. Even simple capabilities such as being able to choose a default font for a word processor presuppose that there exist possibilities to choose from. To design such a space of possibilities, which is meaningful in terms of the work of the users, is no easy task; to make it possible for the users to explore this space is even more difficult.

One major problem is the lack of suitable concepts and "parts" at the levels between a full-fledged application and the programming language. With the older technologies there is usually a gradual movement from the understandable to the incomprehensible. Thus to almost any car driver, "motors" make perfect sense, most understand "spark plugs," and quite a few will even attempt to adjust the distance the spark has to travel. When trying to dig into computer applications, users are almost immediately confronted with concepts and parts that do not make sense in terms of their work. This chapter attempts to begin remedying this situation, trying to set the stage for moving from the computer bound concepts to the work of the users. But there is a long way to go, and the chapter is in some places more technical than the rest of the book.

The rest of the chapter is organized as follows: first we discuss the relationship between use and designing in use and take a brief look at the people doing the tailoring. Then in the main section of the chapter we highlight the practice of designing in use. We look first at how users change the behavior of systems, and then at how to support these activities. Following this is a section on the "down side" of tailoring, on the difficulties that may arise. The chapter concludes with a section summarizing how to support tailoring in the initial design of systems.

Tailoring and Use

Tailoring a system, continuing designing in use, is an activity different from initial design. The activity is related to specific use situations and the result is not a new system, but a modified system; that is, a system with a history which relates it to the earlier version and problems with its use.

While initial design and use thus may be treated as clearly different activities, the distinction between use and tailoring—designing in use—is somewhat blurred. For a particular system a distinction may be made, but in general it is no easy matter, because computer systems, their use and tailoring, differ as much as cars, clocks, and computers differ.

The intuition behind our presentation is coupled to the notion of change, and implicitly coupled to the related concept of stability. The distinction between tailoring and use thus rests on the understood and intended variability of artifacts and their patterns of use. Certain aspects of these artifacts we, as users, regard as stable; others we regard as more or less constantly changing. This relative stability of certain aspects is exactly what allows us to consider tailoring as an activity in itself: we tailor when we change stable aspects of an artifact.

But this stability/change dichotomy is not that crisp, as illustrated by the following examples. I may change the font of a piece of text in a text editor. Is it tailoring? Or how about modifying a spreadsheet? Or moving windows around on the screen to make working on a particular task easier?

Are these tailoring? Or are they "just" use? The text editor is designed for the creation of text in different fonts, so to take advantage of that capability seems like using the text editor, not tailoring it. Similarly, moving windows around on the screen feels more like using the window system for what it is intended than it feels like tailoring it.

We offer two more distinctions to improve understanding of the concepts of tailoring and use. The first of these rests on the view of technology as a tool for addressing certain subject matter, often conceived of as manipulating certain objects. A text editor's subject matter is text (or documents); it manipulates text. A window systems manipulates windows. The distinction then is: if the modifications that are being made are to the subject matter of the tool then we think of it as use; if the modifications are to the tool itself, then it is tailoring. So changing a font on a word in your document is use, while changing the default font—modifying the tool—is tailoring. It should be noted once again that this does not necessarily create a crisp distinction: What is it when you create a new style in your document? That act (at least in Microsoft Word) both creates a style and changes the selected paragraph; so in this distinction, creating a new style can be both use and tailoring. Note also that one person's tool is another person's subject matter, and hence, what is tailoring for one person can be use for another. For the Pascal programmer, changing the compiler is certainly tailoring, but for the team developing or maintaining the compiler, it is use.

The last distinction concerns the duration of the effect of the change. If the modification is so that an effect can be achieved later, then it is tailoring. If the modification is made so that the effect is immediate only, without any impact later, it is use. Thus, if I need a window which is buried and I move windows around on the screen so that I can access it, that is use; if I rearrange windows so that when I next want a project folder it will be easy to find, that is tailoring.

These three distinctions align fairly well in the sense that they seem to get the same answers. Usually, changing the tools is for later effect, and usually, the tools are more stable than the "material" they are used to transform. However, not all these distinctions are meaningful in all cases. The windows of the second example are the subject matter of the window system in both usages; one isn't affecting the window system (the tool) by rearranging windows. (Note that one can affect some window systems: change the layout algorithm from overlapping to tiled, or set it so that it grays out windows which have not been touched for a while.) For this reason, all three distinctions are offered as a set, complementing one another, and jointly informing the concept.

Who Are the Tailors?

After having discussed the activity of tailoring, let us briefly consider the people doing it. Who are they? The simple answer to this question is: those who design in use. In a paper by Trigg, Moran, and Halasz (1987), the discussion is restricted to activities carried out by users. However, the hospital example discussed previously illustrates what we consider to be successful tailoring of a system, but not one done by the users directly. It was done by people from the company that delivered the system in cooperation with one of the users, because the users did not have the programming skills to make such changes on their own.

From our perspective the decisive point is not whether the users actually do it all themselves. We don't assign tailoring to one specific group. The central view is that the change is being made in response to local needs. Obviously a number of changes will be tailored by the users themselves, or not tailored at all, simply because they involve such small details that it makes no sense to involve people outside the use situation. A few examples of this kind were already illustrated. But most cases of more comprehensive tailoring involve some local expert or supporters or application specialists, because the users don't have the competence.

The Practice of Designing in Use

In this section we discuss the various activities involved in modifying computer systems. The viewpoint is that of a user turned designer who is trying to change the way in which technology behaves. It may require that a user brings in others with special skills to help achieve those changes.

We address the practice of designing in use in two parts. First, we present three kinds of activity that change the behavior of the technology. These activities are the normal focus of discussions of modifying technology. Second, we discuss the social and cognitive needs of people engaged in making such changes. These activities are often overlooked, because they do not directly change the technology.

Changing Behavior

Choosing between Alternative Anticipated Behaviors

Many systems allow the users to choose between alternative behaviors that the developers have anticipated will be needed in particular situations of use. The standard mechanism for doing this is through software switches or parameters. An electronic mail system may have multiple delivery services that it can use to deliver mail; a switch is provided to allow the user to choose the one that is appropriate for the local site. Word processors must use some font to display text when the user does not designate one; a parameter is provided which the user can set to indicate which, of all the fonts available locally, is the default font to use.

With this kind of tailoring we utilize a predetermined set of dimensions from which we choose among the different options provided in the same way as we may adjust the front seat of a car. In well-designed systems the possibilities make sense to the users and are easily identified, for example, via menus. For these reasons the users are able to adjust a system by these means. On the other hand, the possible adjustments are limited to those which have been explicitly anticipated by the designers.

In the early use of Buttons, it quickly became clear that for any particular button, the points of change were usually fixed: on the address button, people would change the label and the address (but not the font, look, size, help message, etc.). A mechanism was added to Buttons which permitted button designers to pick out these points of change; that is, parameters were added. These parameters were used to replace general editing with a simpler process of choosing as the means for altering commonly changed aspects of buttons.

Sometimes the range of choices for any particular dimension of variability can be extended by a user through simply adding new material to the environment in which the system runs. Thus, new default font choices are provided by buying and installing additional font definitions.

A limiting case of achieving different behavior from technology is to replace "the whole thing" with a different one. One word processor may provide the footnote format required by a particular journal;

when writing for that journal, a user can achieve the behavior desired by choosing that word processor. In the case of buttons which failed, or failed to meet new challenges, requests to fellow workers and supporters were a common way of getting a new button to simply replace the unsatisfactory one.

Where parameters are concerned, the major lesson for designers is that parameters themselves must be seen as elements of the resulting tailorable system. Thus parameters must be definable, and manipulable in uniform ways. Computer users must be able to find what parameters are there, what behavior they affect, what values are legitimate, etc. The supporters and designers (including, ultimately, the users) must have tools to add new parameters to the collection. This viewing of parameters as tailorable elements in their own right is the direct consequence of our requirement that the tools for tailoring be tailorable themselves.

Constructing New Behaviors from Existing Pieces

Some systems are created as configurations of smaller parts. Construction sets of all kinds exist in non-computer based technologies: office furnishings are composed of pieces of furniture; book shelves are composed of shelving and brackets; Lego and Erector sets create wonderful arrays of toys. In computer systems, file systems and folders mimic the physical world; spreadsheets are composed of linked cells; accounting structures are built of linked spreadsheets.

The behavior of such systems can be changed by changing the configuration. Sometimes the change can be simple rearrangement: in offices we push around the furniture; with desktop computers, we push around the windows. In other cases, more elaborate constructions are required to meet the changing needs of the situation.

Rooms, a Xerox system for constructing separate "screen work spaces," has one such mechanism for constructing rooms from other rooms: It was discovered in making and using rooms that many rooms would share a collection of common tools—the clock, your calendar, an application window for alerting one to the arrival of new mail, a door to get to the mailroom. To accommodate this commonality, rooms were provided with the ability to "include" other rooms, with the idea that the common tools would be put in a single "control panel" room that would be included in all relevant rooms. Subsequently, inclusion was used for other purposes too, suggesting that it is an appropriate mechanism for tailoring (cf. Henderson & Card, 1986, and Card & Henderson, 1987, for information on Rooms).

A special case of construction is that of "accelerators." A system is augmented by constructing a new operator from a sequence of operations grouped together. Ideally, the new operation may be treated in the same way as the original operations, and in particular can be used in the construction of yet larger and more powerful operators. The address buttons discussed above are accelerators. A prime factor in the success of the Unix operating system is the incorporation of such a construction philosophy for commands in the various "shell" command languages.

Another special case of construction as a way of adapting technology is "going outside the system." Here, the system is modified not by rearranging its internal parts, but by using it as a component of a larger constructed system which does what is desired. Thus, to take an example from non-computer based technology, if the range of settings for the driver's seat in our car isn't good enough, we may add a pillow to the seat.

Another argument for using a software "component" in multiple constructions is that the interface need only be learned once. For example, the task of identifying a file on the Macintosh is supported by a single dialogue which is used by most applications. However, the quality of the integration is crucial. Consider the case of spell-checking: When the first spell-checkers arrived for the Macintosh they were separate programs that could process a number of different file formats, including those used by MacWrite and Word. The spell-checkers solved the problem of spell-checking on the Mac in a uniform way, but the integration was so poor that Word users quickly abandoned them when Microsoft offered spell-checking as an integrated part of Word.

A special case of commonality is that involved in the common use of an object (with behavior) as the base for many specializations. Thus, for example, the fact that the generic button supplies a common set of interactions for moving, copying, editing, deleting, inserting in a document, etc., means that all buttons inherit these interactions and need only be learned once. Designing good specialization, in both the mechanisms and the concepts expressed using them, is hard to do well. System designers can provide good mechanisms. They can also provide good examples of how to break up concepts so that specialization is easy and productive. However, at the center of the creation of good conceptual structures is good design and good taste; teaching it may be hard.

The lesson for designers of systems that can be tailored through construction is that the elements and mechanisms which support construction must form a coherent assemblage and support a coherent tailoring activity. This lesson is present but hidden in the term "construction set," which appeals to our sense that sets are the results, not of chance, but of conscious focused design. That is, construction must be designed.

Altering the Artifact

As the most radical kind of tailoring, a last resort when other simpler means fail, we consider altering the artifact itself. (In the case of constructed systems, we consider not the reconstruction, but only the alteration of the smallest nonconstructed pieces.) If we look at the alteration of non-computer based artifacts, we may consider such examples as cutting 25 cm off the legs of the old dining table and painting it, to turn it into a coffee table for a child. Turning to computers, such alteration involves a potential alteration of all parts of a system, particularly the source code.

Changing the source code directly is in most cases infeasible. The code is usually only understandable to the designers; there is no obvious relation between the appearance and behavior of a system and the code. Consequences of specific changes are not detectable in any comprehensive way. Updating and maintenance get extremely complicated. And finally copyright concerns will in most cases make it impossible to get or provide the source code, at least at a reasonable price. Despite these difficulties, changing the source code must be retained as an option to consider.

But means of altering code other than direct change of the source are available. The most powerful of these are code architectures that make it possible to add to the behavior of code without modifying it. A number of such possibilities are found in object-oriented programming; inheritance or specialization, for example. The invocation mechanisms for functions on objects in object-oriented code make it possible to define new classes that specialize the behaviors of existing classes that one cannot examine for the reasons just mentioned. An "event hook" in LISP is another feature that allows users of a system to write LISP functions and install them on objects, to be executed whenever the corresponding event occurs for that object. For example, when the user clicks on a button, the LISP expression which is the value of the "ACTION-hook" is evaluated.

These ways of extending the behavior of code, however, require programming skills. This is the point then at which users may call on supporters for help. It should be noted that there is plenty of precedent in the non-computer world for obtaining such expert help: as the very existence of the profession of tailors makes clear, "users" of clothes do not all possess the skills needed to alter them. When dealing with computers, however, lack of knowledge about possibilities for tailoring may severely limit the users' initiative.

The preceding characterization of tailorable systems is heavily informed by the formative paper on the adapting and tailoring of the NoteCards hypertext system (Trigg et al., 1987). That paper focuses on the characteristics of technology in order to define properties of systems. Here, we focus on the activities that the tailors must do. Our understanding of the relationship between these two different perspectives is as follows. The four properties of adaptable systems are that they should be: 1. *Flexible*: a system with objects that the user can interpret in different ways. We regard such adaptability as addressing not the system but its use—a crucial capability, but not one that we are discussing here. 2. *Parameterized*: offering a range of behaviors from which to choose. We see this as focusing on a particular means of choosing. 3. *Integratable*: a system that can be used in a construction set. Using a system in a construction is only one of the activities associated with construction. 4. *Tailorable*: a system that allows the user to change the behavior of its parts through accelerators, specialization, and adding functionality. We see accelerators as construction, while adding functionality and specializing behavior are altering the artifact. In summary, these characterizations are similar and should be seen as carving up the same space according to different perspectives.

Supporting People in Changing Behavior

We now turn to supporting people engaged in the activities of designing in use. A simple view of these activities is: Some people "author" changes to the technology and make them available; others, perhaps including those authors, make use of the changes. Changes, like the technology itself, become objects to be trafficked in by users. Not only is improved behavior to be had, but membership in the "community of users" may depend on taking part in usage discussions concerning them. The interaction of people with changes becomes merely another case of involvement with design of the system; but, of course, in the case of tailoring, the design will take place locally, motivated by local needs, and often driven or implemented by the users. We can think, then, of a "life cycle" of changes, and look at the support needed to maintain its activities: creating, protecting, perpetuating, sharing, inspecting, learning, and providing feedback.

Creating

The first job is to make authoring as easy as possible. The easiest way of creating something is to copy it from someone else (see *Sharing*). This was the usual way for workers to get a button to do something they wanted. If borrowing fails, the next best move is to copy something you have that is like what you want and modify it. This was the second most common way for workers to acquire buttons. Sharing and copying require finding out what is out there, or nearby; so support for this means of creation is the support of information access: finding, negotiating, making available.

Authoring should require as little new mechanism as possible. That is, provide modification using familiar means. What particular means are already familiar to the user (turned designer/author)? Most immediately, there are the objects and mechanisms of the system being changed. If the changes can be expressed using the system itself, then mastering new mechanisms is not required. For example, a macro in Microsoft Excel is made up of commands. Commands are expressed as cell formulae. Thus, the language of commands in Microsoft Excel is an extension of the language of cell formulae that the user already knows. In contrast, the most general means for changing Buttons is the LISP expression editor, something not required in using Buttons.

Another related source of familiarity is that which a user has with the use of the system itself. This resource can be directly tapped through various means of "construction by demonstration." For example, consider creating composite commands: the user enters a demonstration mode and carries out the sequence that is to be captured; the system watches and creates a new command encapsulating the sequence. Most UNIX shells have this capability.

A third way to tap familiarity is to use the same tailoring tools in many applications. Once a user has learned how to author a change in one part of the system (e.g., in one application), authoring changes in other places will be already understood. For example, the fact that Buttons uses the LISP expression editor means that it is familiar to many LISP users because many applications appeal to this tool. And in contrast, the creation of macros in Microsoft Excel as discussed is completely idiosyncratic, working only in Excel.

Finally, familiarity with the objects of a system can be harnessed to capture regularities by letting the system extract an abstraction from a set of constructed objects, all of which are examples of the pattern (abstraction) sought (cf. Myers (Ed.), 1988).

Protecting

Ease of authoring is not an author's only need. In addition, there is the concern that in changing the system for personal use, a user may mess it up for someone else. This is not a problem if all changes affect only the local author and remain with the local author. But most authors want to share good things, which leverages the effort expended in authoring the change. Thus, for example, when making variable something that had been fixed, one can run the danger of forcing all existing users to learn about and make a new choice. To avoid this problem, an author can insure that the system default to the old behavior. Under this scheme, users need learn nothing about the new choice until they really have a need for it.

And finally, the author must make the change easy for a user to learn about, to learn, and to use. This requires documentation. Further, it is desirable that a user (who may be the author) be able to continue to evolve the change. To aid this continuing design, the history of the design can be a great help, either in conveying the thinking behind the design, or in putting a subsequent author and the original author in contact.

Perpetuating

Changes become a central part of the definition of a system, because they must be added and deleted, set and manipulated, talked about and shared, by the members of the community of users. Thus it is important that means be provided to support changes as objects (objectification of changes) that will be perpetuated while the system remains in use.

First, a change must become an object. Sometimes a change is captured as a procedure to apply to a system (e.g., adjust the time-zone to your local site). These are quickly objectified by creating an object that when applied in some way to the system will carry out this change.

Changes must be objectified for the users, too. A key activity is giving them names or other short descriptions so that they may be referred to verbally and searched for in storage structures. For example, choices in programming environments are often named by the variable in which the value for the choice is held.

Means for storing and applying the changes must be provided. "Initialization files" are often used, either to encapsulate a single change, or to hold a collection of changes. An important requirement for these storage mechanisms is that they support not only individual changes, but the means of assembling them into structures.

One view of buttons is that they are objectified actions. A crucial part of their becoming a powerful mechanism for tailoring in the research communities that use them is the provision for perpetuating them. A means was provided to automatically save buttons such that when a new system was loaded, the buttons in the old one would magically (without user knowledge or action) be restored. This made it worthwhile spending effort on crafting and adjusting buttons.

Sharing

First the news must be published that a change is available. Any or all means of communication can be used. Most interesting are mechanisms that are themselves part of the system. For example, Xerox ViewPoint has a Loader mechanism that can find all enhancements and applications that have been published locally. UNIX has the MAN (manual pages) mechanism that organizes on-line documentation. Buttons were included in documents and thereby passed around as e-mail, and appeared in documentation.

Next, means must be available for acquiring the change. Here again, system-based mechanisms can organize the work, providing uniform strategies for acquiring changes. Finally, taking advantage of changes sometimes must be coordinated among collaborators. For example, if a user unilaterally acquires a new font and employs it in documents, collaborators who lack that font may have difficulty reading the documents. Such coordination requires a user to be able to know who is affected by the change being contemplated and to interact with them. When people do not want to change, it may also require more complex mechanisms for supporting out-of-sync evolution of systems, in which systems with different modifications can "inter-operate" together. For example, most document editors can upgrade a file written by an older version of the editor; fewer— but some—are prepared to read the newer (and non-understood) objects in a file created by a newer version, preserve them across editing, and write a newer version file.

The other side of this requirement is the need to help users stay current. As colleagues evolve their use of shared systems, they may find that they are operating in a way that is regarded as obsolete and outdated. Keeping completely current may be very time consuming. Knowledge of what others are doing is needed to support achieving the necessary balance between change and stability. Thus, sharing must include knowledge not only of available changes, but also of current usage patterns and understandings of acceptable practice in the local community.

A special case of coordinating with others is the establishment of the necessary assumptions upon which a change depends. Buttons created severe problems for workers on more than one occasion when e-mail containing buttons was not even readable because the receiver's environment lacked something required for the button to display itself. Eventually, button mechanisms were designed so that a failure did not disrupt the receiver's system. The lesson for the designers of tailorable systems is that the architecture of the mechanisms must be robust enough to handle errors in the practice of tailoring. Handling errors means not only minimizing failures, but also providing enough information to diagnose and report failures that do occur.

Inspecting

A user wanting to change a system will want to know what changes are already there and their history. This history includes the changes that have been made, who made them, and for what purpose, including the work practices that the users had in mind when modifying the system. All of this is needed for people to be able to continue to evolve the system coherently. For example, my Macintosh may be set up to mount certain directories automatically on the local fileservers; some directories are mounted to support working on certain projects, others are a function of people, others of social conventions of the working groups; others using my system casually may find those arrangements quite unclear.

Learning

Reading documentation and talking to people about using a system change is one thing; learning to actually use the system in its modified state is quite another. Real learning requires much additional work, both to find out how the technology actually works in detail and, more importantly, to learn how to use those capabilities in their work. This involves the mapping of task needs to system capabilities and the development of practices of work to support the use of the system. For example, making copies of large documents on certain copiers requires that there be a flat working surface to hold intermediate stacks of paper, and a large stapler, with a desk strong enough to take the pressures of using it. Buttons are easy to use (just push them), but the effect is not always so easy to predict. The central Buttons mechanism includes a means of giving a user simple help. In addition, conventions have developed encouraging the publishing of buttons in documentation, preferably in e-mail. These conventions support an important pattern (see *Feedback*) in which trouble with use can be reported by simply replying to the e-mail that brought the button to you. However, once received and put onto the desktop (put to work), buttons lose their ability to lead the user to documentation or the author.

Such learning must inevitably take place in an exploratory mode. Some attempts will fail badly. Users will therefore want an environment for supporting this learning with two important properties:

First, it must be possible to experiment in a safe way, in a way that guarantees that no damage will result from misunderstood changes to "real" information. For example, users experimenting with Excel often open a new worksheet to try out new formulae and macros, knowing that changes usually happen to the sheet in which the formula is located. Second, after exploration has settled down, "real" work may have been done in that safe but "unreal" environment which one would then like to preserve. For example, copying the newly debugged formula from a trial Excel spreadsheet to the real one is quite easy.

Providing Feedback

The final step, closing the development cycle, is to provide feedback to the creator(s) of changes, based on experience with using the modified system. This is essential for the continued tracking of user needs by the authors of changes. When a user is the author, this activity amounts to the user reflecting, on occasion, on experiences with the changes that have been created. Interestingly, the detection and reporting is often considerably easier in the case of difficulties than for success. Difficulties make themselves known by their negative results, to which the user is usually forced to attend; successes are often very much harder to see; something going right is usually the result of many things going right. And, there is the human question of motivation: "If it ain't broke, don't fix it" is the death of much feedback.

To provide feedback requires knowing what information would be useful, who to feed it back to, and how to get it there. This is the other half of sharing, and the considerations on information dissemination apply here just as they did there. Mechanisms for providing feedback can be built into the system itself. For example, the installation procedures on the Hermes message system ensured that any user with a comment could simply use the "comment" command and Hermes would get the comment to the developers (Henderson & Myers, 1977; Myers & Mooers, 1976). The user needed to know nothing at all about who the developer was. And with buttons, particularly those that came in e-mail, when they didn't work, one could reply to the e-mail (thereby getting it back to the sender without thought), including a few well-chosen words about what had gone wrong—a very easy exchange indeed. The lesson here is that tailoring should not be considered independently of the communication mechanisms that will support sharing and feedback.

More important than the mechanisms for delivering experiences of use to the authors of change is the establishment of a climate of collaboration among users and authors that fosters the learning that all use makes possible. There is a cost even to sending a comment, and more cost if the system has been very difficult to use and the user can't characterize what happened or went wrong. The investment in feedback, however, can be seen—if the larger social system is correctly constituted—as not only a duty, but as a likely source of a better tomorrow.

Some Problems with Tailoring—Why Not Tailor?

Before we conclude by summing up how to design for tailoring, we look at the down side of tailoring, at some of the major pressures against changing the behavior of a system.

Tailorability Is Hard to Provide

As discussed previously, to make tailoring possible may require implementing lots of switches, construction sets, mechanisms for adding behavior to code, and integrating into communication systems and social systems. This is not always a large job, but it always adds to the work and in some cases is a great deal of work indeed.

Second, the mechanisms for allowing design in use must be provided. An interface (or interfaces) must be designed and built. It is interesting to note that while buttons have been in use for a number of years, the mechanisms for editing them are still under development.

Third, mechanisms for saving and reestablishing the state of the system must be built. The mechanisms for appropriately saving and reestablishing can be difficult to construct. No current system does this in a principled way, the problem being to untangle the tailoring done by the possibly many authors who have provided changes to an environment, once it is in place. There is research needed on this issue; the principles have not yet been worked out. In Buttons, this difficulty manifests itself in not knowing, for a button which has passed through many hands, whom to give feedback to; or when a new version arrives in the mail, whether the new one should be used in place of one the receiver already has and is happily using.

Finally, there is the very pragmatic fact that all these additional mechanisms add to the resources necessary to use the system. Effort must be expended to create them, adding to the cost of acquisition (construction expense or price). And the system is normally larger, requiring more machine resources to run it (memory and disk space).

Learning

Turning to the user side, a system with more "stuff" in it requires more learning. A user must learn about the various design possibilities and what aspects of the system's behavior they affect. When the means for design are programming language expressions, the language, its forms and their meanings, must be learned. The interface(s) to the tailoring mechanisms must be learned too, including, where relevant, editors for creating and manipulating linguistic expressions.

Beyond learning the details of the technology of tailoring, a user will need to understand why a particular possibility has been provided by a designer. That is, what applications are better supported by the system if it behaves in one fashion or another. Consider, for example, what happened when word processors were upgraded to permit user control over the fonts in which their text appeared. People usually had no training, skills, or even taste in the arena of typographic design. In many places there was a severe outbreak of "font-itis," the inappropriate use of fonts in documents.

Finally, systems used together often must have their tailoring coordinated. Consequently, a user may well have to learn how colleagues and correspondents tailor their systems for the collaboration to work. For example, if one user employs a package of macros for the spreadsheet program Excel, those sharing these spreadsheets will have to do the same.

A variant of the learning task is the remembering task. Once committed to using certain features of a system, there can be a continuing difficulty in making use of that capability. One forgets how to use it and has to turn to manuals, examples, and colleagues to reproduce the knowledge.

Using

If one wants to or is forced to tailor, a lot of time may be required. In particular, a changed system will inevitably require changing one's practices of use. It is a very common feeling that one may never regain in time saved, the time spent learning to save it. When the mechanisms to protect against unsuccessful tailoring (breaks, wrong function) themselves are unsuccessful, this can be doubly annoying and time-consuming. Difficulties in saving and restoring buttons have on more than one occasion severely endangered their continued use in one research lab.

Tailorability produces an explosion in configurations in single systems, and when systems interact (when don't they?) the possibilities are endless. A plethora of configurations increases the likelihood of interactions that have to be dealt with by tinkering until things work. Careful factoring of effects can help manage this complexity, but trying out a new configuration may waste a great deal of time and try a lot of patience. So the users will bear the cost of possible tailoring while they are using their systems, whether they tailor them or not.

Finally, but often foremost, is the possibility that the variability provided is just not of any interest to the users. The fact that multiple fonts are available may not be of interest when writing a book which will be typeset by the editors or publishers. The provision of many features, so called "creeping feature-itis," is an ever-present source of unlooked-for and unwanted tailorability, for which users pay a price.

Babel

A common experience with tailorable systems is to sit down at a colleague's machine that runs systems that you use on your machine and discover that you can't use them because they have been tailored in ways that you are unfamiliar with. Keyboard "wirings" and macros are particularly difficult to adjust to because one's patterns of usage get "built into one's synapses"—backspace is in a certain place (the delete key, or control a, or leftarrow) and the impulse to backspace produces finger motion faster than one can easily control. In Buttons, the unavailability of buttons familiar on your own machine may make it impossible to do things you very commonly do "at home."

In some situations, it would be useful if one could install one's own personal "style," including packages, "over the top of" your colleagues' behavior. This is not usually easy to do, because the "saving technology" is not sophisticated enough to save the personal style separately from everything else. For all these reasons, people are loathe to adjust other peoples' systems to their own styles. This makes sharing machines harder when tailoring is common.

Of course, there are cases where you would not want to change to your personal style anyway. When you are trying to help figure out a problem, particularly concerning the behavior affected by personal tailoring, or when you are working collaboratively with a colleague with both of you seated at her machine, you are depending on her style being in force. However, with extensive tailoring, the unfamiliarity of styles—indeed, the languages of interaction—can greatly increase the difficulty of even talking with one another about the things that happen in the system. It becomes harder to help or debug, and collaboration can be reduced to content, not implementation. Other familiar cases are naming files or arranging directories: when collaborating using files, personal styles conflict, and compromises or decisions to live with incompatibility and inconsistency must be made.

The Missing Designers

Finally, in those cases where professional designers are absent from the "tailoring loop," tailoring may even decrease the usefulness of a system. The subtitle of this book emphasizes the cooperative nature of good design, and just as we have argued for the need to include users in the design process, the competence of designers will often be needed. Particularly in cases where the professional competence of the users does not otherwise include tailoring, it may be wise to seek assistance before modifying a system. An example of this kind is described in Grudin and Barnard (1985), in which command names tailored by naive text editing users were much less effective than those supplied by skilled designers.

Designing for Tailorability

Tailorability is no panacea, as the preceding discussion illustrates. And what is more, possibilities for tailoring will in many cases be application specific, requiring genuine design work by the creators of a system. However, the reasons for tailoring remain, and experience indicates that the resources saved in the initial development by ignoring tailorability issues will be spent several times over during the use of a system, either by the people doing the maintenance, or by the users who have to take still more elaborate action to bend the system to changing circumstances.

The lesson is design and architecture. If an artifact is created with change in mind, even if the specific change is not anticipated by the designer, the creation of change can be greatly simplified, regularized, and most importantly, supported. A simple example of this was given early in this chapter, in which new help messages were added to a system.

An important step in providing more user-accessible tailoring will be to develop new high level, or application-oriented, building blocks together with application-oriented ways of manipulating them. In terms of our characterizations in this chapter, altering of artifacts will be transformed into construction by finding decompositions of those artifacts that align with the understanding of users. The third case described in this chapter, Buttons, is a first step in this direction.

Object oriented design and implementation are a promising basis for this kind of work, since they may allow us to shorten or even bridge the gap between the structure of computer systems and their code on the one hand, and the structure of the work and concepts of the application area on the other.

Acknowledgments

Thanks to Jonathan Grudin for tailoring our language and to him and many of the authors of this book for helpful comments.

References

Brad A. M. (Ed.). (1988). *Creating user interfaces by demonstration.* San Diego, CA: Academic Press.

Card, S. K. & Henderson, A. (1987). A multiple, virtual-workspace interface to support user task switching. *Proceedings of CHI+GI '87,* (pp. 53-59). New York: ACM Press.

Grudin, J. & Barnard, P. (1985, April). When does an abbreviation become a word? and related questions. In *Proceedings CHI '85, Human factors in computing systems (San Francisco),* (pp. 121-125). New York: ACM Press.

Henderson, A. & Card, S. K. (1986). Rooms: The use of multiple virtual workspaces to reduce space contention in a window-based graphical user interface. *ACM Transactions on Graphics, 5* (3), 211-243.

Henderson, D. A. & Myers, T. H. (1977). Issues in message technology. In *Proceedings of the Fifth ACM/IEEE Data Communications Symposium, September 27-29, 1977* (pp. 6/1-9). New York: Snowbird, UT, IEEE.

MacLean, A., Carter, K., Lövstrand, L., & Moran, T. (1990). User-tailorable systems: Pressing the issue with Buttons. In *Proceedings of CHI '90* (pp. 175-182). New York: ACM Press.

Myers, T. H. & Mooers, C. D. (1976). *HERMES user's guide.* Cambridge, MA: Bolt Beranek and Newman Inc.

Sheil, B. (1983, February). Power tools for programmers. *Datamation, 29* (2), 131-144.

Trigg, R. H., Moran, T. P., & Halasz, F. G. (1987, September). Adaptability and tailorability in NoteCards. In *Proceedings of INTERACT '87.* Stuttgart, Germany.

12

From System Descriptions
to Scripts for Action

Pelle Ehn and Dan Sjögren

Carpenters at work.

Prologue—"Woodlands of Scandinavia"

We are in the late 1970s somewhere in the woodlands of Scandinavia. A group of carpenters are sitting in the lunch room of a carpentry shop. A die is rolled. A token is moved on the board and hits a "market opportunity." One of the carpenters picks up a "market card" and reads: "The big furniture company is offering you an order of 200,000 shelves of their book-shelf called *Captain*.

They will pay you 10 kronor per piece. You must have at least 20 'capacity points' in your factory to take the order. Your production will be tied up for three months. You will not be able to take any new orders during the next three turns." The carpenter uses a minute for reflection, interrupted by comments from his co-players. He says: "Ok I have chosen to play the role of big business owner." He takes the order, borrows the money from the bank, invests, and installs a new line in his factory. He will not get paid until time for delivery. Big risks, and big profits. But will he survive during the next three turns at the die?

The carpenters are playing Carpentrypoly, a game similar to Monopoly. But the group of carpenters playing Carpentrypoly is doing serious work in a local "study circle." They want to find out what kinds of consequences different business strategies will have for the design of technology and organization, and the impact on quality of the product and the work.

The game was used in a development context, a trade union-based educational program for members in small and medium sized carpentry shops in Sweden. The goals of "Our Shop," the educational program, were to strengthen the positive aspects of the workplaces such as good product quality and good workmanship, and improve the weak sides, such as bad organization and poor physical environment. The educational program was meant to lead to action, to proposals and demands for improvements in factory and workstation design. The educational program had the form of a local study circle, which is a common practice in Scandinavian trade union education. The overall program was conducted by a shop steward supported by researchers and industrial designers. The basic work, however, was carried out in local study circles where carpenters from the shops in question together with the researchers and the industrial designers engaged in a mutual learning process— designers learned about the carpenters' situated work and carpenters learned about design methods.

In developing an *action program* for changes to the workplaces, the study circles used three design games. In addition to Carpentrypoly, we developed and used the Layout Kit and the Specification Game.

The Layout Kit consists of a collection of cards of common wood working machines and accessories. The carpenters used large sheets of paper on which a layout of the factory walls had been drawn. To create a common vision and a shared understanding of the present state of the shops, they placed "machine cards" on the sheets. The cards were used to lay out existing shops and to identify problems. They were also used to design new alternatives both for the shop lay out in general and to individual workstations.

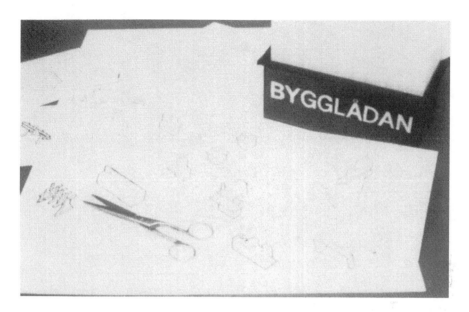

The Layout Kit consists of a number of cards, each with an icon of a machine.

In Carpentrypoly, the carpenters were given an opportunity they never had in real life. Everyone inherited a worn-out factory from their parents. The participants could choose among three typical roles reflecting different business ideas and strategies: The tycoon type, the stingy one who hated to put even a dime in "unnecessary improvements," and a third role, the enlightened owner with a reasonable respect for the qualifications of the workers and a sober attitude to technology, preferring intermediate solutions. The board reflected a year, during which the players got market opportunities through market cards. They could also respond to changes in the economic and political context in society. After the game, the outcome was assessed and discussed, focusing on the relations between quality issues, business ideas, and technology and work.

The results from investigating the factory layout and market relations are later structured and refined using the Specification Game. The views of products and production that the owner had when he inherited the factory and the changes that had taken place since then are placed under the headlines of "Product," "Technology," "Organization," and "Work." Through the debate about the quality issues, the workers formulate their own demands. The outcome of the Specification Game was situated demands on "Product," "Technology," "Organization," and "Work."

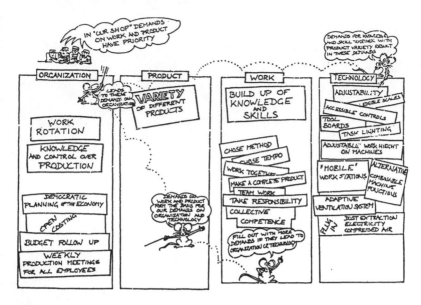

The Specification Game—the headlines and the demands from the workers' point of view.

After the Specification Game the study circle returns to the Layout Kit to outline the ideal design of their "own" factory layout according to the demands they have formulated.

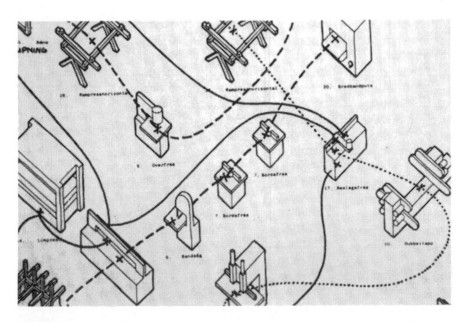

Production design using the Layout Kit.

The participants also select the one workstation that they think most urgently needs redesign. Again, they used the Layout Kit, but now it is in order to rebuild a single workstation. Different solutions to the problems can rapidly and easily be laid out and debated.

When the new layout has been made, the Specification Game is played again. Demands on the products, the technology, the work, and the organization are scrutinized in relation to their ideal layout.

In the *local action program* of the carpenters, all demands and proposals are prioritized, argued for, and scheduled under the head-lines of "improvement in the working environment and construction of new workstations," "improvements in the work organization" and "improvements in the design of products and extensions of the pro-duct range."

After the study circle, some of the design proposals from the ac-tion program are built as mock-ups or prototypes, to provide hands-on experience before actual investment and implementation.

A prototype of a low cost multi-functional and adjustable worksta-tion using powered hand-tools.

The Social Construction of System Descriptions

Serious Design Games?

Are the "low tech" trade union carpentry production activities discussed in the prologue relevant to the high tech area of system design? Can playing really be taken as a serious design activity? To us there is a positive answer to both questions, answers that we elaborate on in this chapter.

The trade union study circle Our Shop and the design artifacts used by carpenters in the woodlands of Scandinavia to develop the quality of work and technology had their origin in the Carpentry Shop project, a research and development project we participated in in the late 1970s and early 1980s (Ehn & Sjögren, 1986). Active user participation and improving the quality of work and products were seen to be main factors in supporting democratization of the work place, which was the overall goal of this and similar research projects, such as the DEMOS project, that we were engaged in at that time (Ehn, 1989).

Our design experiences with the Carpentry Shop project were important when we moved on to design integrated computer-based text and image processing systems for newspaper production in the UTOPIA project (see Chapter 7). Though the application domains and level of technology were very different in the two projects, they had many features in common.

Some central features were the participatory design approach and the understanding of the design process as a process of mutual learning between professional designers and skilled users within the application domain, and as a process where future or alternative technology and work organization were envisioned and experienced rather than described. Aspects shared by the design approaches included a focus on concreteness and ease of use. The design approaches included mock-up simulations, prototyping, and organizational games supporting investigatory work in study circles and in design groups. The use of mock-ups and prototypes opened up possibilities for "design-by-doing"—for getting hands-on experience with future technological alternatives (see Chapters 9 and 10). In this chapter we focus on the use of organizational design games and how they may create opportunities for "design-by-playing," involving participants in design discussions of the overall work organization, skill requirements, division of labor, and cooperation in the work process.

The idea of design-by-playing is not really a Scandinavian invention. For example, we could have learned something similar from Russell Ackoff and his practice with social system design (Ackoff,

1974). He concludes that for participation in design to be successful it requires that:

- it makes a difference for the participants
- implementation of the results are likely
- it is fun to participate.

The two first points concern the political side of participation in design; the users must have a guarantee that their design efforts are taken seriously. The last point concerns the design process. No matter how much influence participation may afford, it has to transcend the boredom of traditional design meetings to really support design as meaningful and involved action. According to Ackoff, the principles of *idealized design* lead to a design process that is both liberating and fun. The design work is treated as play. The approach we had taken has a family resemblance with this participatory idealized design.

We did, however, learn more directly from the workers at LIP, the clock factory in Besançon, France. When faced with the threat of unemployment in 1974, they occupied the factory and kept on producing clocks, but they also made alternative products. One of these products was the direct inspiration for Carpentrypoly. This was the strategic Monopoly-like game Chomageopoly (an unemployment game). To them unemployment was the hard reality, serious enough for the game of Chomageopoly. The design games we suggest in this chapter are to be used in the serious realities of participatory design. These games support engaged and even pleasurable involvement, not only for the fun of it, but exactly because it takes the actors' concerns seriously. This is also why we, as explained in the next section, were forced to take the step from system descriptions to scripts for action. In the third section we illustrate how development of scripts for action can benefit from deliberate use of games and dramatic play metaphors. The last section, the epilogue, goes beyond design-by-playing, making it part of a *Future Workshop*.

The Breakdown of System Descriptions

As system designers we are familiar with making and using system descriptions similar to the one shown below in Figure 1. Some prefer working with object-oriented description methods like *JSD* (Jackson, 1983), others prefer data flow-oriented descriptions like *SA/SD* (DeMarco, 1978). Being from Scandinavia, we have been trained in the similar methodology called *ISAC* (Lundeberg, Goldkul, & Nilsson, 1978).

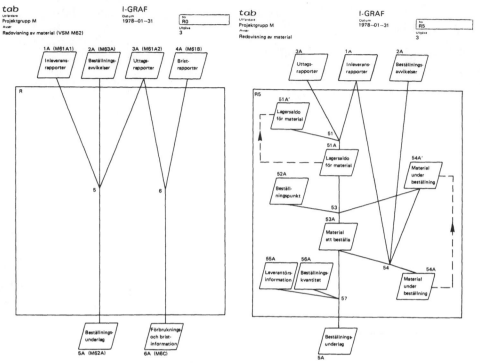

Figure 1. A system description.

This was our approach in the initial analysis and design work in the UTOPIA project. As design experts we tried to capture the participating typographers' views of work organization and technology, and of production and information flow. At first we thought that we were very successful with our many system descriptions. As designers we quite liked the systems we were designing, and we thought that the typographers were pleased with the descriptions as well. There came a day, however, that put an end to this idyllic form of designer-user cooperation. This was the day when we found out that the system descriptions only made sense to us, the system designers (see also Chapter 7). The only sense our system descriptions made to the participating typographers was that they were made by us, that is, their own experts. There was no co-design going on. Our system descriptions did not support user participation; to the users they were literally nonsense. We had to re-think our use of design artifacts for descriptions if we wanted to do participatory design. No matter how suitable the traditional description methods may have been for specifying requirements for technical implementation, they did not support participatory and involved acting. Hence, the move away from the safe ground of correct system descriptions to the open-ended games of participatory action.

One of our first moves was to develop a new design game. We called it the Organizational Kit and the idea was to enrich abstract system descriptions work with the participatory lessons we had learned in making and using the situated Layout Kit for carpentry production.

As with the Layout Kit for carpentry production, the basic ideas behind the organizational design game in the UTOPIA project were that:

- it should be fast and easy for a group of people to work with;
- it should be cheap and flexible to use, allowing several alternatives to be tested during discussions;
- it should be based on concepts relevant to the actual type of production and support design discussions of existing and future work and technology.

Reconstructing Newspaper Technology and Work

The following pictures are from a design game played with the first version of the Organizational Kit for newspaper production. The focus is not on the product as in traditional system description methods, but on the design process as a social construction of reality (Berger & Luckmann, 1966). The basic idea is to support involved discussion about design alternatives as seen from the perspective of changes in work and work relations.

To support this *the semantics and syntax of the game have to be situated.* Hence, in contrast to traditional description methods, we do not think of objects and relations as general concepts, but as specific to the domain. There are, for example, *specific* functions in a *specific* work process, such as "kerning" (adjustment of space between letters to increase legibility) in page make up, specific tools like "laser printers," and specific materials like "articles," and "display ads." These are depicted graphically by icons that make sense to the participating users.

In the UTOPIA project, the Organizational Kit for newspaper production was developed and used as follows.

We set out first to find a useful metaphor for the whole game. Lack of imagination made us decide on a traditional production flow metaphor. Then we tried to define the basic functions in this production flow, and the artifacts and materials that might be used. We defined about 40 such functions (e.g., page make up, page layout, registration, editing), 15 basic artifacts (e.g., scanner, desktop laser printer, pen), and five types of material (e.g., article, page, logotype). For each function, artifact, and material a card type picturing an icon was made.

UTOPIAn designers and users reconstructing their understanding of reality using the Organizational Kit.

The cards had different colors (white for functions, yellow for artifacts, and red, orange, pink, blue, and green for materials) and different sizes. They were all adhesive. The cards were duplicated and organized in a box; they should be easy to move around on the common playground, a piece of wallpaper.

Function cards, artifact cards, and material cards.

By placing cards and discussing their meaning, a shared understanding of the current organization with its problems was to be supported, as well as ideas for alternative organizational and technical designs.

As a background for this initial construction, we as designers conducted interviews, discussed with users, and "hung around" at relevant work sites. We also read technical reports and domain specific material about newspaper work and production.

In the next step the users were introduced to the rules of the game. Typically this was done by using a concrete example. Our paradigm cases were centered around two newspapers we had studied, the *Star News* from Pasadena, outside Los Angeles, which had fired their typographers when computer-based page make up and picture processing were implemented, and *Østlændingen* from rural Norway, which, although almost equally advanced technically, had found an organizational model for peaceful coexistence between typographers and journalists.

A typographer and a journalist working together in the new work organization at the newspaper Østlændingen in Norway.

After this introduction the users started to use the kit, placing and moving the paste-up cards around on the wallpaper. The process typically started with the design group making a description of the users' existing newspaper. The design group then moved on to discuss changes and develop alternatives. In this process where both users and designers were actively involved, new functions and changed rules were introduced when they made sense to all participants. The descriptions were often "humanized" by making and placing icons for human actors and professions in relation to the functions. This made it possible to discuss qualifications and educational demands as well as work environment problems and solutions. The final design, not only of solutions, but also of the semantics and syntax of the game, was in the hands of the playing participants.

Playing organizational games with the Organizational Kit is basically a learning process. For all participants in a design group it serves as a means to create a common language, to discuss the existing reality, to investigate future visions, and to make requirement specifications on aspects of work organization, technology, and education. These have been our main uses of the game. Our UTOPIAn organizational design game has, however, also been used in other situations of social construction of reality.

For example, we were invited to play the game with representatives from the Newspaper Employers Federation and from the Typographers Trade Union as preparation for national negotiations on new technology in the newspaper industry in Denmark. The "Scandinavian model" for work organization and new technology in newspaper production that was developed through participatory design using the kit is now codified in many technology agreements in Scandinavia. (For further details on the game and on how it has been used see Dilschmann, Ehn, & Sjögren, 1985.)

But are these kinds of semistructured design games better for participatory design than traditional description methods for structured analysis? We think so, not only from practical experience, but also from a theoretical perspective. This perspective is founded in Ludwig Wittgenstein's language game philosophy (Wittgenstein, 1953).

What a Picture Describes is Determined by Its Use

In an approach to design inspired by Wittgenstein (Ehn, 1989; see also Chapters 7 and 9), the focus is not on the "correctness" of system descriptions in design, on how well they mirror the desires in the mind of the users, or on how "correctly" they describe existing and future artifacts and their use. System descriptions are seen as *design artifacts*; typically, linguistic artifacts. The crucial question is how we *use* them, what role they play in a design language game. The new orientation suggested by a 'Wittgensteinian approach' is that we see these design artifacts as a special kind of artifact that we refer to as "typical examples" or "paradigm cases" when we describe something, or when we "inform" each other. In the language game of design we use these artifacts as *reminders* for our reflections about future computer artifacts and their use. The use of design artifacts brings earlier experiences to our mind and "bends" our way of thinking of the past and the future. Thus, we should understand them as *re*presentations. If they are good design artifacts, they are useful in supporting good moves within a specific design language game, but this is not a question of correct mapping. We have to look for design artifacts that *make sense to all partici-pants*. Design artifacts such as those used in the organizational design games are, in our experience, good examples. Other good examples focusing on the practical use of future computer artifacts are the mock-ups and prototypes discussed in Chapters 9 and 10.

Creating New Design Language Games

As designers we are involved in reforming practice; in our field this typically means creating or modifying computer artifacts and the way people use them. Hence, the language games of design change the rules for other language games, those of artifact use. What are the conditions for this interplay and change? What do we as de-signers have to *do* to qualify as participants in the language games of users? What do users have to learn to qualify as participants in the language game of design? And what means can we develop in design to facilitate these learning processes?

To possess the competence required to participate in a language game requires a lot of learning within that practice. In the beginning however, all you can understand is what you have already under-stood in another language game. You understand because of the *family resemblance* between the two language games. However, by understanding design as a process of *creating new language games* that have a family resemblance with the language games of both users and designers, we have an orientation for doing design as skill based participation. This approach offers a way of playing and do-ing design that may help us transcend some of the limits of formal system descriptions, making it possible for users and designers to put the "tacit dimension" of their practical social and instrumental skill into play.

Viewed this way, the design games from the woodlands of Scan-dinavia and from our UTOPIAn newspaper production may by useful paradigm examples and reminders for other participatory designers to try out in reality. But is there not more to be utilized in the family resemblance with traditional games? This was a question we asked ourselves when we entered the office environment for a game of desktop publishing.

Playing in Reality

From Production Flow to Social Construction

Certainly many different activities are going on in offices. Here we will concentrate on changes related to the introduction of desktop publishing.

If we compare the design game we developed for desktop publishing with the design games from the UTOPIA project and the Carpentry Shop project, there are not only similarities, but also clear changes. The participatory language game approach is still fundamental. Concreteness and ease of use are still basic principles. However, the flow oriented production metaphor used in the earlier games did not make sense in the office environment of desktop publishing. Hence, the metaphor shifted toward *a game based on linguistic actions leading to commitments* (Winograd & Flores, 1986). Social interaction and cooperation come into focus more than the physical artifacts used (see Chapter 3 and Wynn, 1979). Furthermore, the *play metaphor* is used much more explicitly and the theoretical motivation for this is elaborated. We have transcended the Monopoly-like board game approach from the Carpentry project in one important aspect. The idea of the game is neither to encourage competition nor to teach a theory from above, but to support situated and shared action and reflection.

Another striking difference is the design focus. Existing hardware and software are more or less taken for granted. This is based on the assumption that the basic problem in the domain is not technology driven, but a question of organizational change, education, and requalification. The existing technology may be good enough. The focus for the professional designer shifts from software development toward utilizing available hardware and software while considering work organization and product quality.

A Serious Game of Desktop Publishing

We have played the Desktop Publishing Game in a few public administration offices in Sweden. Our main case, however, is the consumers agency Konsumentverket. The dramatic design context the game offered at Konsumentverket was based on six concepts:

The *playground* is a subjective and negotiated interpretation of the work organization in question. The *professional roles* are represented by both individual professional ambitions and organizational requirements. The *situation cards* introduce prototypical examples of breakdown situations. *Commitments* are made by individual role players as actions related to a situation card. *Conditions* for these commitments are negotiated, and an action *plan* for negotiations with the surrounding organization is formulated. These concepts were used throughout four development steps.

The game started with the Prologue, a half day session where the rules of the game were introduced, the proposed playground was redesigned, and the cast was chosen. The First Act was a two-day session in which, based on situation cards, situations were simulated and commitments made under certain conditions. The Second Act, another two-day session, was based on an updated playground where the work with real publications was simulated. In the Third Act the game was turned back to reality and an action plan for negotiation with the surrounding organization was formulated. The use of these concepts and the structure of the game will be explained as the story of the game at Konsumentverket is being told.

The Desktop Publishing Game has material for making *graphic reminders* like notice boards, cards, and personal scripts. It is *language* oriented in its use of professional and situated concepts from relevant professional traditions and from the application domain. It is *concrete* in its use of critical situations and events in the actual organization. It is *dramatic* not only in its use of the game metaphor, but more specifically in its use of a theatrical play metaphor.

To facilitate the reading of our paradigm case, the basic concepts and the development steps are summarized in the figures below.

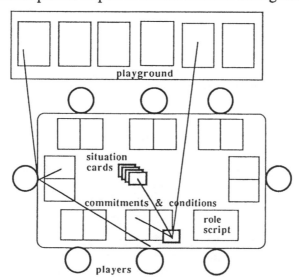

Figure 2. Basic concepts in the play situation.

In the play situation shown in Figure 2 a situation card is drawn by a player. Given her professional role as identified by her script, she makes a commitment under certain conditions that are negotiated with the other players. These commitments and conditions are written down in her script and are also placed in a situation on the playground. Other players, too, may make commitments and state conditions to help resolve the situation triggered by the situation card.

Design-by-Playing at Konsumentverket

Konsumentverket, the national Swedish board for consumer policies, has an educational department which deals with information for consumers. The 25 employees at the department produce books, reports, educational material, and brochures. The department had invested in desktop publishing technology. However, the lack of familiarity with this technology caused uncertainty about professional roles and an inability to raise the typographic quality. To really use the technology required development of professional roles, change in professional relations, education, and a redesign of work organization. This was the situation when, in the early spring of 1988, we responded to a request from management and began a project financed by the National Fund for Administrative Development and Training for Government Employees.

Setting the Stage

At the first meeting between us, the designers, and the participants from Konsumentverket, we introduced the idea of the game as a way of developing professional roles and restructuring the work organization. The basic rules of the game were explained using examples from another case. The participants from Konsumentverket described problems they had in their practice and showed examples of products they had made. This was the first step in establishing the shared language game to be played by us and eight employees in the education department of Konsumentverket. They were secretaries, case handlers, a technical support person, a marketing person, and the head of the department. An information consultant and a typographer also occasionally participated as resource persons.

The Prologue

When the stage was set, we started to work on the *prologue*, which meant preparing the *playground* and the *cast*.

The playground was a crude reminder of the general work tasks that publication work normally consists of, and was to be used as a kind of bulletin board for commitments and conditions. It was made up of twelve sheets of paper, each referring to a general work task. These work tasks ranged from publication ideas to marketing. The tasks were drawn from typographic and editorial professional traditions.

A playground with the work tasks from "the idea of a publication" to "the marketing of a publication."

At the prologue meeting the playground was put on the wall. We discussed the work tasks and how they were labeled. The playground was revised and clarified until it worked as a reminder for the participants' situated experience and until consensus was reached on the interpretation. For example, the work task "investigation" was replaced by "gathering of facts."

The cast was prepared for traditional roles such as the editor, the executive editor, the graphic designer, and the author. But there were also new roles such as the technology coordinator, the publication administrator, and the "formatting person." These were extensions of the secretarial role. The roles were partly taken from the situated practice at Konsumentverket and partly from roles we had encountered or created in an earlier case. For each role a *script* was prepared.

At the prologue meeting, however, the cast was also revised. For example, the participants proposed that the role of "picture editor" be integrated into the role of the editor, and, after discussion a "marketing" role was added. The role scripts were distributed and studied by the participants, each of whom chose at least one role. The roles were chosen according to realistic professional ambitions and organizational demands. For example, one secretary chose to play the role of the "formatting person" and another chose the role of the "publication administrator."

Each script consisted of the following parts: qualifications, tools, and work tasks. The actors prepared for Act I by studying and revising their scripts, and by writing situation cards for the plot.

The situation cards propelled the plot forward. Each card reminded the participants of a critical event, a breakdown situation, or a daily triviality which sooner or later was likely to occur. There were three types of situation cards: paradigm cases from other organizations using desktop publishing technology, role oriented cards, and cards specific for the actual organization. The organization specific cards on problems at the education department had been prepared by the actors; the others were prepared by the designers.

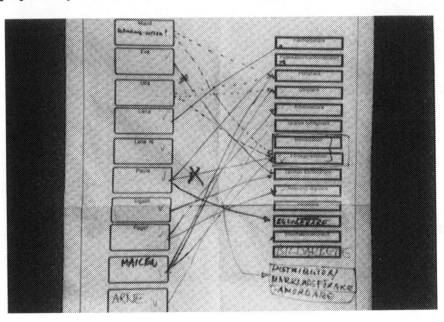

The role cast prepared and proposed by the designers and revised and changed by the actors.

Act I

The actors met with the designers on the stage. The playground was placed on the wall. A total of 81 situation cards were put in a stack on the table. The actors had put their individual role scripts on the table. The designers gave a short introduction to Act I. The play could begin. One of the participants picked a card, which read:

Printing of camera ready original on laser printer is still in progress. It must be delivered to the print shop before 4 p.m. today. The toner cartridge is suddenly empty, and no one has bothered to order new ones. Who takes responsibility to solve the problem now? And what should be done to avoid similar breakdowns in the future?

This certainly reminded the participants of a familiar situation that typically "falls between the cracks." After a discussion the secretary who had chosen the role of the publication administrator made a *commitment*. This commitment meant that she took responsibility for solving the problem under certain *conditions*. For example, she stated:

In my role as publication administrator I commit myself to solve the immediate problem by ordering a new cartridge and have it delivered by taxi. In the long run I will establish a routine for the maintenance of the laser printer. My condition is that I get appropriate training in maintenance of the printer.

The group actively supported her commitment, which was then written down on a situation card. The card was placed on the playground in the work task labeled "printing." She made a similar note in her own script.

The game continued, with more cards handled in the same way. The group at Konsumentverket got on well and had no problems with the cards, even though some situations were tricky. However, in some situations the participants agreed that the problem was irrelevant. If no commitment or agreement had been reached in a situation, the designers were prepared to intervene to try to re-interpret the situation to reach a consensus. Such an intervention did not, however, occur during this game.

The second day started with a review and a discussion of an updated playground and the new scripts. The rest of the day the group worked with the remaining situation cards in the stack. To get some variation and greater involvement, the group split up into smaller groups.

By the end of the second day, Act II was prepared. The group was then asked to select three different publications that were typical of their present production. At Konsumentverket the group chose a catalogue with a complicated layout, a book, and a newsletter. As preparation for the next act, both the participants and designers wrote five new situation cards based on their experience in working with these publications.

Act II

A new gathering on the stage. The scene had changed. Now the play came closer to the situated work at Konsumentverket. The new cards reminded the participants of earlier breakdown situations. However, the task in the game was now to "replay" the chosen publications in the light of the roles, commitments, and conditions agreed upon in Act I.

During the first day the playground was adjusted for each of the publications and the new situation cards were played. At Konsumentverket this turned out to be much more complicated for the participants than the work in Act I. For example, one of the new cards on "the catalogue" read:

The catalogue was scheduled for the annual education fair, but due to production problems the catalogue was delivered too late to be printed in time for presentation at the fair.

This situation card led to an animated discussion of who was to blame, rather than commitments on how to solve the problem given the new roles.

Given this breakdown, the designers decided to introduce a change in the rules of the game as the game went along. Hence, on the second day, a new type of card labeled "development card" was introduced by the designers. The development cards were formulated as a response to problems that arose during the "replay" of the publications. For example, the development card that read:

How can we improve proofreading of our publications?

was generated in response to a qualification and a task that seemed to be underestimated in the organization. The discussion of the card led to an acknowledgment of this problem, a person who committed herself to solve it, and an agreement from the group that this was an essential task.

By the end of Act II about 150 commitments and conditions had been formulated and agreed upon. As preparation for Act III, the designers updated the playground with these agreements.

Act III

From play to reality was the theme of the last act. In Act III the players prepared to meet their audience, and earlier commitments and conditions were tied together in an action plan for negotiations with the surrounding organization.

As designers, we introduced six headings for refining the commitments and the conditions into demands in the action plan.

The demands concerned:

- *publication program*—how to develop the products
- *professional support*—ideas and proposals for competence which must be "imported" into the organization, such as typographic competence
- *development of qualifications* within professional roles
- *technology support* for assessment, investments, and the maintenance of the technology
- *organizational support* concerning relations to the surrounding organization and management
- *internal work practices*—new routines, quality control stations, and teamwork that the participants can implement themselves.

This structure for the action plan was based on the designers' earlier experiences with making action plans with users. The structure was a suggestion to the participants, a suggestion that could be changed and modified. At Konsumentverket, however, the participants accepted the six headings.

The action plan at Konsumentverket in progress. Formulated commitments and conditions from the play are refined under titles such as "development of publications" and "development of qualifications" (in Swedish).

For all work situations on the playground, commitments and conditions were cut out and pasted into the action plan under the appropriate heading. For example in the work situation "formatting," the technical support person had committed herself to update all desktop publishing software continuously if she were offered appropriate training and time to do the job. This commitment was pasted under the headings "technology support" and "development of qualifications."

The last day was dedicated to scrutinizing and refining the demands in the *action plan*. All commitments and conditions under each demand heading were evaluated, assigned a priority, and debated. Only demands for which there were a group consensus were kept in the action plan. For the five first headings these demands were the group's *requirement specifications* to Konsumentverket for changes. For example, the group demanded that proper training on desktop publishing software and software support be part of the job description of the technical support person.

The last heading, "internal work practices," was treated differently, since these were demands that the group could implement without external support. They "only" needed consensus in the group.

When the playground had been transformed into a script for action, it contained 38 suggestions for improvements, all of which had been debated, evaluated, and refined during the game.

Nachspiel

The official evaluation report by Konsumentverket summarized the experiences with the game in the following way:

> The project has made an open discussion of everyday situations possible, sorted out misunderstandings related to work tasks, and pointed out difficulties in the publication process. The educational department has strengthened its ability to collectively and directly solve (professional) problems. Analyses, problem-solving, and education have been integrated. The work roles and the potentials for developing them have become clearer. The development project has led to a new "work style" which the participants can argue for and which they can teach colleagues who were not involved in the project. The new work style is clearer and has improved internal and external communicative competence. (Konsumentverket, 1988, p.5)

From the perspectives of the participants and the official evaluation, it seems that the costs related to the game were offset by improvements in practice and by reductions in time-consuming and cost intensive breakdowns.

It is, however, an undisputable fact that the game was time-consuming the way it was played at Konsumentverket. Today, however, we, as designers, are able to carry through a game like the one at Konsumentverket in approximately half the time. One reason is simply that we have become more experienced and are better able to improve and change a game as we play. We have learned to be more supportive of and instructive to the participants, while knowing when to stop interactions that are not clearly focused, and to skip parts of the game when appropriate, and to adopt to the situation.

This chapter started with study circle attempts to democratize and improve working conditions in carpentry shops. Then we introduced our UTOPIAn organizational game on quality of work and product in design of newspaper technology. Finally, a game of work relations and new technology in the office environment was played. This may look like a shift in focus, away from serious democratization, toward superficial games for enjoyment. Has shop floor reality been substituted for the creation of illusive games? This is an important question and a serious criticism, but we think it misses the main point. The shift from system descriptions to scripts for action has been venturesome but necessary, in order to take the users' everyday practice seriously, making them the main actors in participatory design.

In retrospect, the construction and use of the playgrounds seem to be the least successful part of the games we have been playing. There seems to be a risk that conservatism will support a traditional production flow oriented view of work and technology. To overcome this is certainly a challenge for future games, as is the shift back from the the focus on linguistic actions and commitments in the office to the more tangible material world of shop floor production. In the epilogue to this chapter we will take this challenge seriously and suggest a future game to be played in shop floor reality.

But before leaving the game in an office environment, we would like to end with a critical situation card that must be better understood; an understanding that may change the scripts for participatory action:

> Who will win the desktop publishing game, the manager or the secretary? Is design-by-playing a game of seduction rather than one of liberation?

Epilogue—Proposal for a Future Game

In this concluding scenario we will return not to the woodlands of Scandinavia, but to the old industrial belt in mid-Sweden. It was in this setting that back in the 1970s we carried out the DEMOS project with the State Railway locomotive engine repair shop in Örebro (Ehn, 1989). The game described here has not happened. We present it as a possible next step in combining techniques for design as action.

Locomotive repair workers in action.

In the 1970s a major problem for the workers and their local union was the threat of ISA-KLAR, a computer-based planning system. The system, with automatic work orders that would deskill workers and control their work in detail, represented management's alternative to piecework wages, which the union had gotten rid of after a long struggle. The local union formed an investigation group that together with designers from DEMOS developed alternative plans for work organization, production planning, and a well-functioning work place. As a result of the workers' investigations, ISA-KLAR was never implemented. However, neither were the workers' organizational and technological alternatives.

In addition to wages, democratization of the work place and good product quality are now as then the main issues for the union, and production planning remains one of the major problems in the repair shop. However, a major change in attitudes has occurred. As summarized in an evaluation report: It is not an exaggeration to say that the management philosophy on efficient organization has now caught up with the one the union developed in the mid 1970s (Brulin, Mehlman, & Ullstad, 1988, p. 34). Hence, the repair shop in Örebro seems to us to be an ideal place for a Future Game.

Main participants in this design game would be the repair workers from the shop, with production management on different levels also taking part. We do not think this contradicts democratization, given the strength of the union and the new management attitudes.

Contrary to our point of departure in the DEMOS project, we now think that linguistic actions and commitments are as important aspects of work on the shop floor as they are in the office environment. Hence, we claim that it is just as relevant to play a linguistic game in the repair shop as it was in the case of Konsumentverket. Actually, we think this may be useful for overcoming the spell of the traditional production flow approach to shop floor work.

We also have to admit that some workers were bored by the design methods we used in the 1970s in Örebro, preferring to stay by their machines rather than join the design meetings of the union group. We think of involvement in design games as a way of overcoming this boredom, making design more engaging and hopefully pleasurable.

We suggest that the proposed design game in Örebro should be integrated with a Future Workshop (see Chapters 8 and 13). We think this will open up possibilities for a participatory way of developing playgrounds and situation cards that are genuinely situated and relevant to the participants. We also think that the use of design games will make Future Workshops more dynamic and creative in fantasy development and in making implementation plans.

A Future Game at the Locomotive Repair Shop

We imagine the following situation in the locomotive engine repair shop in Örebro: Production planning is regarded as a major problem by both workers and management. The uncertain deliveries of repaired locomotives due to delays of tools and materials create a lot of problems for the repair workers as well as for management. A skilled work force is seen by both management and the union as a necessity for quality production. Computer support for material administration has again been suggested as a remedy to the problems of production planning.

This is an outline of a possible Future Game to be played:

The Preparation phase. As designers and game facilitators, we would meet with management and the local union to discuss the Future Workshop and to initiate negotiations for participatory conditions. We would also "revisit" the shop floor to listen to old friends' and new acquaintances' views on production problems.

A Future Game with the title "Computer support for material administration?" is agreed upon, and undercarriage repair work is cho-

sen as a test site. An invitation to all employees involved in this area
would then be distributed.

The Critique phase. Eighteen employees (15 repair workers, two
technicians, and the production manager), together with designers as
game facilitators, constitute the design group. Initially, the workshop follows the traditional pattern, with participants making short
critical statements, that are later prioritized and grouped into themes.
These negative themes would be converted into positive ones.
These could, for example, be "long term planning," "development
of qualifications," and "autonomous repair teams."

After this, however, a game would be constructed. The positive
themes are the material from which a playground is made by the
game facilitators. Based on the critical statements, all participants
make specific situation cards. A card could read, for example: "You
are mounting the wheel axles of an undercarriage, but the tools you
have ordered are missing."

The Fantasy phase. It is here in the Fantasy phase that the game
would be played. Situation cards would be drawn, commitments
made, and conditions stated. Every participant would act according
to their professional ambitions. Since this is a fantasy, "unrealistic"
commitments and conditions are accepted.

Before starting to play, the design group would split up into three
teams. Based on the games, the teams continue working on their
fantasy, and prepare a presentation for the rest of the group.

As a response to the situation card suggested above, a player
could make the commitment that necessary tools are available when
needed by the team. As conditions for this he would demand investment in a complete set of tools for each repair team, and a budget for each team to replace worn out tools.

As in other Future Workshops the fantasy themes would be presented, discussed, and prioritized.

Based on the prioritized fantasies a new playground is made by
the game facilitators, and all participants are asked to make new realistic situation cards related to these fantasies.

The Implementation phase. The Implementation phase would start
with the participants playing the new game. Again they would
divide into three teams; this time, however, they play with real commitments and realistic conditions. For example, on the question of
available tools, a commitment may be made under the condition of
investment in more tools and better routines for keeping track of
them.

After the game the participants meet in a plenary session where
commitments and conditions are edited and prioritized as in the case
of making an action plan in the game at Konsumentverket.

Following Future Workshop practice the participants would decide how to initialize the first steps in the action plan. One such activity is to initialize concrete negotiations between management at the repair shop and the local union concerning project groups and their goals and resources; another is to disseminate the experiences from the Future Game into discussions on the shop floor.

An old photo from the locomotive repair shop. Will it also be a future workshop for participatory design games?

In the design process that follows this Future Game, more games may be played, and mock-ups and/or prototypes experienced. There will certainly also be a time and a place for design artifacts such as detailed technical system descriptions, but that is another story. The story we have been telling about scripts for action is about the participatory side of design, and the necessity of taking users' experience seriously. That is why we have been playing games, not at the price of seriousness, but as a necessary precondition for engaged and more democratic participation.

References

Ackoff, R. L. (1974). *Redesigning the future—A system approach to societal programs.* New York: John Wiley.

Berger, P. & Luckmann, T. (1966). *The social construction of reality—A treatise in the sociology of knowledge.* New York: Doubleday.

Brulin, G., Mehlman, M., & Ullstad, C. (1988). *På rätt spår.* Stockholm, Sweden: Arbetsmiljöfonden.

DeMarco, T. (1978). *Structured analysis and system specification.* Englewood Cliffs, NJ: Prentice-Hall.

Dilschmann, A., Ehn, P., & Sjögren, D. (1985). *Gränslandet.* Stockholm, Sweden: Swedish Center for Working Life.

Ehn, P. & Sjögren, D. (1986). Typographers and carpenters as designers. In *Proceedings of skill-based automation.* Karlsruhe, Germany.

Ehn, P. (1989). *Work-oriented design of computer artifacts.* Falköping, Sweden: Arbetslivscentrum and Hillsdale, NJ: Lawrence Erlbaum Associates.

Jackson, M. (1983). *System Development.* Englewood Cliffs, NJ: Prentice-Hall.

Jungk, R. & Müllert, N. (1981). *Zukunftwerkstätten—Wege zur Wiederbelebung der Demokratie.* Hamburg, Germany: Hoffman & Campe.

Konsumentverket. (1988). *Slutrapport för projekt Dnr 39/88 hos Statens Förnyelsefonder, Dnr 88/A37.* Stockholm: Konsumentverket.

Lundeberg, M., Goldkul, G., & Nilsson, A. (1978). *Systemering.* Lund, Sweden: Studentlitteratur.

Winograd, T. & Flores, F. (1986). *Understanding computers and cognition—A new foundation for design.* Norwood, NJ: Ablex.

Wittgenstein, L. (1953). *Philosophical Investigations.* Oxford, UK: Basil Blackwell.

Wynn, E. (1979). *Office conversations as an information medium.* Berkeley, CA: University of California.

13

Epilogue: Design by Doing

Joan Greenbaum and Morten Kyng

*[This] divorce of art from technology is completely unnatural.
It's just that it's gone on so long you have to be an archaeologist
to find out where the two separated.*

Pirsig, 1974, p. 148

Some have said that ours is a holistic approach, something like
Pirsig's story in *Zen and the Art of Motorcycle Maintenance.*
"Holistic ideas are fine," they argue, but they usually go on to say
that system development is really a science, and after all, "We can't
tackle the whole, without starting with individual problems or
parts." Others who have read our work or talked with us grumble
over the fact that the ideas presented in this book are fundamentally
political because they start from the perspective of people at work,
of the workers. For them this approach is interesting but a bit too
idealistic, and their arguments usually sound something like, "Well,
in the real world, you know, it's the managers who allocate the
money for the systems." And a third and vocal contingent of critics
complain that we set out some intriguing ideas, but don't really explain how to "do" them.

In August of 1989 the contributors to this book held a two-day
workshop to sort through what we thought we said and to look for a
way to say what we really wanted to. Never an easy process, and
this one was made particularly difficult by the fact that as sixteen
authors sat with two and three drafts of chapters piled around their
feet, a warm, sunny, August breeze swept through the windows.
To propel ourselves out of our summer lethargy we decided to hold
our own "condensed" future workshop, spending a half day on the

project. The theme we chose was "Cooperative Design—Year 2000." The group looked at where they would like to be in ten years; what another book might look like then; what they would hope to achieve in that decade and how to cooperate to do it. The idea was that the authors would go through the three stages of a Future Workshop (see Chapter 8), voicing our frustrations about what we hadn't gotten to do (the Critique phase), giving shape to our visions about what we ideally wanted to do (the Fantasy phase), and then, with a sharper focus, coming back to sit down to proposals for how to get going, including hard-nosed proposals for editing (the Implementation phase). Later in this Epilogue we tell some stories from that workshop, as a way of saying that we recognize the critics' voices among our own, and have been actively engaged in a dialogue to make this book as useful to others as we possibly can.

Like designers of a good computer system, the authors had to stop, take a breath and reevaluate the ongoing process. And to improve the usefulness of the product we too cooperated with the users in the design process. To that end, selected potential users have read different versions of the chapters and discussed changes with the authors and the editors. And, as in the design of computer systems, substantial improvements have been made.

Most of our readers, however, will be in the same position as the typical users of "off the shelf" software; they are users who had nothing to do with its design. Like such users, our readers may feel that their needs and interests have not been as much in focus as they would have liked. In some respects our readers are even worse off, since the book has none of the tailoring possibilities of good systems! In the Preface we appealed to the imagination of our readers;

here we supplement that appeal by some of our internal dialogue. We tell about the process of designing the book: what we set out to do, what changes it went through, where we hope to go with it, and the rationale behind it.

On the drawing board, this book was to be called "Design by Doing." "Doing" was a central concept for us, embodying the idea that there needs to be active involvement of users and designers working together at activities that actually "do" something. As John Dewey (1938) explains in *Learning by Experience*, learners shouldn't be "spectators" or passive participants in the learning process. Designing and using a system is certainly a learning process for all groups involved, requiring that we all do something about it. But at the same time several of the authors felt somewhat uncomfortable with the title. As we began to discuss it, the original clarity quickly evaporated. In what sense isn't traditional system development "doing?" and "How do intellectuals 'do'?" were some of the questions raised in our Future Workshop. Perhaps in our eagerness to get "doing" back into the picture, we were not just balancing the scales, but introducing a new, false dichotomy between doing and reflecting?

So midway through our writing process we shifted our focus slightly to acknowledge that what we were actually doing was showing ways to reflect on work practice and how to give that practice a central position in the design process. In Part I, Reflecting on Work Practice, we tried to take the reader through some terrain from the social sciences and humanities to expand our ways of looking at the workplace. Here we discussed the rich tapestry of tacit knowledge in the workplace (Chapters 3 and 5) and pointed to the importance of the web of situated actions in which people find themselves (Chapter 4). We pointed to how learning about the specific cultures of the working groups can make us more sensitive to what kinds of systems people may be comfortable with and therefore use more effectively (Chapter 6). Reflections on work practice, we believe, are critically important for ongoing design, not as laboratory experiments that measure the statistical significance of a user's interaction with a system (Chapter 2), but for daily or routine project work. However, a lot of work remains. The analytical approaches, with their emphasis on observation, listening, and watching, have to be developed further to suit a cooperative design process where the "objects of analysis" stop being objects and instead become active participants.

Part II, Designing for Work Practice, illustrates, among other things, how metaphors are a powerful way to reframe thinking, to understand the present better and push design beyond a single problem-solution approach (Chapter 8). In this part of the book we showed how mock-ups are useful, playful, and fairly cheap as ways to envision future working environments in the early stages of design (Chapter 9). We also explained how cooperative prototyping goes beyond simple demonstration models, giving users hands-on experience with envisioning their own ideas (Chapter 10). Throughout this section we put forward our theoretical starting points, partially rooted in the philosophies of Wittgenstein and Heidegger, that remind us that the way we live in the world, the ways we see it, and the ways we are able to express it, influence what is possible in design (Chapter 7), reemphasizing that we firmly believe that the design process must start within the use situation and from within the experience of the people who will be using the computer system. We also demonstrated that design should be an ongoing process, if systems are to continue to suit their users. To support this, the possibilities for tailoring as an ongoing process need to be considered in the design of a system (Chapter 11). Since organizations change and adapt to new systems, the design process needs tools for working within the changing organization, and users—and designers—need techniques for consciously addressing the questions of enhancing the use of new systems, and of improving the organization of work (Chapter 12).

In reflecting on the contents of the book, we are fairly pleased that it represents a break from the linear, solution-driven engineering approach that has dominated research on system development for the past 25 years. And in reflecting metaphorically on our own writing process, we find that it also breaks away to some extent from that linear, solution-driven approach. While we don't often sound our own horn, we do sincerely believe that the mix of art and science that was stirred up in this volume shows that the development of computer systems for the workplace can be approached in new and effective ways. Yet there are many things left out: ideas and practices that we would like to have expressed, but lacking sufficient experience, both in the "doing" and in the writing, we were not able to do so adequately. To tell you a little more about our frustrations we offer some scenes from our Future Workshop.

A Warm Summer Afternoon

The room is a university seminar room, a bit stuffy, and packed, with tables pushed to one side, coffee cups and papers on every conceivable surface, and sixteen authors-cum-designers, plus the energy of the five-year old son of one of us. We have just finished tossing colored marking pens to each other, as one-by-one we hopped up to scribble words on poster paper taped on the blackboards and walls. Here are some of the things that are taped up there:

Criticism

- no direct contribution from users
- we need a prototype of what we are doing
- where's the beef?
- no aesthetics of design
- many practices missing
- too academic
- cooperation between the practicing reflectioner and the reflecting practitioner?
- we don't have a lot of good products (that is, systems) to show
- too many false dichotomies, e.g., theory vs. practice
- how about the videos?
- we have no traveling road show

Many of us begin to pace around the room, as Morten Kyng, acting as facilitator of the Future Workshop, asks us to vote on what we see as the most important problems with the book. Groans and complaints are heard, as is frequently the case at the end of the Critique phase, when everything that's been said and written seems both so important and so trivial at the same time, that it would be impossible to classify it. But we press on, and after the five-year old has finished drawing stick figures on our works of art, the voting is finished and we are able to group the "winning" complaints around four common themes. These are:

- our frustration with the diverse theoretical base and our inexperience with expressing it in practice,
- the lack of more cooperative relations to potential "users" of our work;
- the lack of "good products" and of recommendations for turning good designs into good products/systems; and
- the lack of empirical projects in real life settings to try out the ideas presented in the book.

The authors feel that only the last three can be worked on at this point, and subsequently these are turned into positive formulations to be worked on in the Fantasy phase. The first of these, dealing with more cooperative relations to potential "users" of the book gets the label "The Traveling Road Show."

It's time for a lunch break, and we divide up into groups, taking our colored markers and poster paper out into the sun to try dreaming in a new environment.

After lunch and a one and a half hour working group session we return to our stuffy room and begin to tell about and act out some of our fantasies of what we would like to do instead of, or in addition to, writing the book. The story telling from the Fantasy phase, like most such workshops, takes on a rather romantic aura as we give voice to our fantasies without worrying about the more practical details ahead. Two of the groups come in with drawings (inspired by the five-year old?) and proceed to talk about their sketches.

One group's story goes like this:

The all embracing real life project: This is a scenario of how a fantasy project might transpire. Imagine that in a municipality somewhere some forward-thinking people read the book, find it compelling, get in contact with us to discuss how we might help them explore the ideas in the book, and we dispatch the Traveling Road Show!

What's going on in this setting? Let's suppose it's a small administration for a local community. The people work well together and are quite satisfied with their work situation except that they have a resource problem—they have lots of ideas for new projects, but like most local governments they don't have the funds to hire more people to do what they would like. So they feel a strong need for redesigning their work to allow themselves to implement more of these visions and increase coordination among the projects they are now doing, which may well involve improving the computer support. At the moment they have management approval to start a project in conjunction with us, and they even have some capital funds allocated for equipment. Together we agree to start a project with a three to five year time frame. The overall goal is to increase local resources for planning and implementing programs. But quick results, say within the first year, are considered crucial for the continued enthusiasm of the participants as well as for management support. In short, we want them to be able to address their immediate problems and at the same time be able to continue working with more complex, long range issues after we've left.

Now to get things going the Traveling Road Show moves in. We try to identify quick-fix possibilities and in parallel begin studies of current work practices and of relations with the community. We arrange visits to similar workplaces that might provide prototypes for redesign and we provide demos of different potentially relevant technologies. Future Workshops are arranged, and eventually two redesign projects that address technology and work organization together are agreed upon.

Another Fantasy phase work group, more interested in products, began their scenario differently:

The Beautiful, Exciting and Suitable Products Department: In our design organization there is a BES department. Its main purpose is to help set up creative environments wherever the roadshow goes. Their slogan is to stimulate imagination, experimentation, and social interaction.

When the roadshow arrives at a new place, the BES department starts its work by visiting the local real estate brokers, where they rent a house in the vicinity of the user organization. Since the house is to become the home base of the designers and working space for users, this is no easy matter. The house must have nice white walls, lots of windows and light, a beautiful garden, and a porch where designers and users can sit in the evenings .... The house is stuffed full of equipment so that ideas springing from casual conversations can be quickly tried out. The group considers this house to be their best resource since experience shows that beauty and comfort in the work surroundings tend to result in beautiful and useful products.

Besides the policy of comfort and beauty, the BES has a second cornerstone in its design approach: waste. Designers and users are encouraged to experiment and throw away; so most of the programs made are tried out and then thrown out, and only a minor percentage live to be products in the end. In addition to the products a collection of program building blocks, metaphorically referred to as the Lego blocks, have been designed, used, and developed further. It is precisely this large stock of "Lego blocks" that prevents the BES from going bankrupt.

The major part of the work is not done at the BES house though, but in prototyping nurseries. Physically the nurseries are set up at the workplace of the users—a couple of desks, with workstations connected to the BES house. Here, in the borderland between design and use, informal sessions are held when users want to try out a prototype, and here redesign and programming take place. In the nurseries both users and designers are in an environment with some family resemblance: the designers know the setup with the workstations, but the users know the workplace.

And so it goes. As system designers we laugh at our simplistic and idyllic visions, and as authors, we begin to wonder how we'll ever express, in written form, even a small percentage of the hope and energy that we see when users and designers work intensely together. The workshop then goes on to the more mundane business of formulating some concrete plans for making the problems less troublesome and the visions a little more probable. By the end of the afternoon, we agree on individual and group rewriting assignments, and procedures for involving potential readers, including a one semester course the following Fall. The Traveling Road Show is left as an intriguing idea. However, the other scenarios in fact inspired later projects: one with a Danish public organization, involved in redesigning their computer support and work practices, and one project based on the BES Lego blocks—but that's another story! What was left at the Future Workshop was to spend a hectic hour trying to clean up our clutter accumulated during the day (the Fantasy phase is fun, but like the Critique phase, it tends to be rather messy). As readers and users of this book, you, of course, are the judges of how well our joint discussions worked. But while we are aware of the limitations of what we wrote about, we are equally aware of some of the things we left out of the book. The Workshop helped us clarify this. Here are some of our concrete complaints and plans for the future.

Some Missing Pieces

For some of us, we wish that we could have focussed more on how work could be made more cooperative, and on how issues of quality of work and democracy in the workplace could be brought to life. The book *Computers and Democracy* (Bjerknes, Ehn, & Kyng, 1987) was essential background for many of us, and the issues of democracy that it raised have been swept to the side here as we focused on how to foster cooperation between designers and users. While we didn't explicitly address concepts of democracy in the workplace, it is clear to us that the issue of building cooperation is a first step on that path. For us, democracy in the workplace means expanding choices for workers and developing working strategies that allow more voices to be heard. In the political arena this means a form of pluralism where many groups learn to work together and struggle with compromise and consensus. Within workplace walls this pluralism, or multiple interest group interaction, is a new and largely ignored phenomena. An emphasis on cooperation, we believe, is a necessary building block in this process.

Another issue that ties in with the first is our concern with the possibilities for developing working groups within the political constraints of organizational settings. Dreams about cooperation and democracy have little possibility for survival if the designers and users don't face the political realities of the companies that they work for. While this was touched on in several chapters, we want to emphasize that building cooperation means building communities of different categories of workers and managers who learn about each others' interests and develop workable strategies for their working environment. What works in a large bank may be totally inappropriate in a government office, and the forms that cooperation takes in a small service firm may not work in a factory.

A significant issue concerning power politics is the role of the designer in "exposing" the tacit skill and workplace culture of the workers. Many issues are involved, including the difference between the more passive stance of researchers examining a workplace versus the activist role of designers involved in changing work practices through computer system design. This issue, like the others we have raised, can't be solved with a simple formula for participation. For us, a rule of thumb has been to remember that things said to us by groups and individuals are said in confidence. Like lawyers, we feel that it's our clients' or users' prerogative to use the information we have gathered with them.

The problem that we call transcendence versus tradition is another fly in the ointment of smooth cooperation. Our dream is that cooperation between computer users and designers can help us transcend or break away from some of the stifling designs of the past. But while designers tend to repeat design procedures from the past, users also tend to be conservative or traditional in how they look at their workplace. Often designers complain that asking users to imagine new forms of work is an impossible task, for given the day-to-day reality of work, most of us see it as it is. The chapters in Part II dealing with what we call envisionment tools offer some ways out of this dilemma, but we are aware that cooperation and envisionment techniques only provide sparks to begin the process of reflecting and acting on technological support for new working environments.

While we didn't intend to write a "how to" book, we are sometimes frustrated by the fact that in this volume we have given you a fair number of examples of what we did, with very little guidance on how we did it. In early drafts, in fact, most authors wrote about what they did and gave suggestions for doing it. But we found that this approach was too general, and that, for the most part, people reading the early drafts still felt confused about how to go out and do it. Indeed if we take our own writing seriously, we repeatedly point out that describing work practice is a difficult and risky venture and that choosing the language game to communicate it in is a matter between the users and designers themselves.

Like Pirsig, whom we quoted at the beginning of the chapter, we find the separation of art from technology unnatural. We are, indeed, trying to put art back into the technology, and in doing so, as with any form of art, one must experience it in order to interpret it in the immediate situation. This is not to say that there are no general guidelines. There are indeed guidelines for this type of experiential work, just as there are guidelines for working with experiential and collaborative-based teaching and techniques for developing art. But these guidelines will have to be developed in further work, work better suited for cooperative learning—perhaps even the traveling road shows that we fantasized about, or the video we mentioned in the Preface.

The Next Frontier

This book began with a short history of a Scandinavian approach to system development. Its roots lay in twenty years of experiments with applying ideas about social democracy to the design of technical support in the workplace. It seems to us that the time is ripe to apply these ideas on a broader scale. American businesses, with their eyes on Japan, have at least been giving lip service to the idea of using team-based management strategies. In some American firms, ideas about cooperative work, group work, and involving users in technical and organizational decisions have begun to be heard. Whatever the reason for starting this process, applying cooperative design strategies is no longer a remote possibility limited to only a few "progressive" firms. In Western Europe, as many companies gear up to a more unified Common Market, they are finding that competition on price alone is ineffective, while international competition based on specialized quality may open new doors. The market for specialized and quality software systems that are tailored to individual workplaces seems a likely possibility in this new era. And in Eastern Europe the opportunity for using cooperative design strategies may fit into workplaces where work groups and cooperatives, of sorts, have already been set up.

There are other trends that we believe support an environment for trying out the ideas in this book. The proliferation of desktop personal computers and local area networks has significantly decentralized computing in the last decade. In addition, the massive market for off-the-shelf software has meant that more and more systems are customized or adapted versions of existing software. Time saved in software development can make more time available for working with cooperative prototypes and taking a serious look at working practices. Increasingly, users, particularly professionals who use computers at home, favor software that can be adapted to their own work practices. And for a majority of workers who have been raised in a computer literate era, the question is no longer "Why did the computer do this?" but rather "Why *can't* the computer system do this for me?" User expectations are rising and ways of tapping their energy and potential need to be exploited.

Democracy in the political arena has been a significant watchword for the last two hundred years. Throughout the Western world, democracy within the workplace is clearly the next frontier. Work life represents one-third to one-half of the daily hours of most adults. And virtually all estimates indicate that the majority of workers will be using some form of computer system during the 1990s. Employing democratic means to design technology seems an eminently reasonable idea for beginning and carrying out workplace democracy. Clearly the field of system development needs to adapt to this changing environment. Cooperative design of computer systems is a good place to start.

References

Bjerknes, G., Ehn, P., & Kyng, M. (Eds.). (1987). *Computers and democracy—a Scandinavian challenge*. Aldershot, UK: Avebury.

Dewey, J. (1938). *Experience and education*. New York: MacMillan.

Pirsig, R. (1974). *Zen and the art of motorcycle maintenance—An inquiry into values*. New York: Bantam Books.

About the Authors

Peter Bøgh Andersen, linguist by education, is associate professor at the Institute of Information and Media Science, Aarhus University, Denmark. The institute provides an interdisciplinary education in computer sciences and humanities that he has been deeply involved in creating. He has a background as a hard-boiled formalist and champion of artificial intelligence but abandoned that field in favor of a semiotic approach ten years ago. His main contribution to the computer field is the introduction of the media perspective as a way of looking at computer systems in use. He has published several articles on the topic and has recently finished a book called *A Theory of Computer Semiotics* (Cambridge University Press, 1990). Beside that he is the father of two children and a devoted reader of science fiction.

Liam Bannon was Visiting Associate Professor in Computer Science at Aarhus University during 1988-90. He is currently engaged in teaching, research and consulting activities in Dublin, Ireland. His background is in cognitive psychology and computing, specifically human-computer interaction, computer mediated communication, and computer support for cooperative work. Previously he has worked with Honeywell, Inc. at their research center in Minneapolis, Minnesota, and has been a consultant on the FAST programme of the European Community. He was also on the human-machine interaction project at the University of California, San Diego, from

1982-85. He is co-editor, with Z. Pylyshyn, of *Perspectives on the Computer Revolution* (New Jersey: Ablex, 1989) and with U. Barry and O. Holst of *Information Technology: Impact on the Way of Life* (Dublin: Tycooly International, 1982).

Keld Bødker is an assistant professor at the Computer Science Department, Roskilde University Centre, Denmark. He teaches system development and does research on theoretical and methodological issues of development of information systems. His Ph.D. thesis dealt with the application of a cultural perspective on organizations and work processes to analysis and design of information systems. He lives with his wife and two children in Copenhagen.

Susanne Bødker is assistant professor at the Computer Science Department, Aarhus University. Her research focuses on human *use* of computer technology and on the design processes connected to this use. Her work includes the book *Through the Interface—A Human Activity Approach to User Interface Design* (Lawrence Erlbaum Associates, 1990), in which she provides a theoretical framework for understanding (and designing) the role of computer applications in use. Susanne Bødker teaches system development and human computer interaction. She lives in a collective community in Aarhus.

Pelle Ehn works at the Institute of Information and Media Science, Aarhus University, Denmark, where he teaches contextual software design and does research on design as cooperative work and pleasurable engagement. He has a background in Computer Science and in Sociology and was for many years a researcher at the interdisciplinary Swedish Center for Working Life, heavily engaged in emancipatory projects such as DEMOS and UTOPIA. He is the author and editor of several books, including *Work-Oriented Design of Computer Artifacts* (Lawrence Erlbaum Associates, 1989), and *Computers and Democracy* (Avebury, 1987) with Gro Bjerknes and Morten Kyng.

Joan Greenbaum, professor of Computer Information Systems at La Guardia Community College, City University of New York, has been a frequent visitor to the Computer Science Department in Aarhus, where she and Morten Kyng taught courses that led to this book. Her first book, *In the Name of Efficiency* (Temple Press, 1979) focused on the history of divided labor in the computer field. Many of the projects she has worked on concern the effects of technological change on women workers. She has been involved for almost two decades in developing active learning strategies to help people learn about technology and economics in order to deal more effectively with issues in the workplace. Joan Greenbaum has three sons and currently lives in Montclair, New Jersey.

Kaj Grønbæk is an assistant professor at the Computer Science Department, Aarhus University, Denmark. He teaches system development and does research on techniques and tools for the cooperative design of computer systems. In particular, he has undertaken several field studies on end-user involvement in system development and how it can be facilitated with different prototyping techniques and tools. Some of these projects were carried out as action research in which he participated in cooperative design activities with different groups of users. Together with his wife and a son, he shares a house in Aarhus, Denmark, with another family.

Austin Henderson is an area manager at the Xerox Palo Alto Research Center in California. He does research on improving the design process through improved coupling of those involved in the development process, particularly users and designers. His past work on providing technology that is modifiable by its users was done in part at XeroxPARC's EuroPARC laboratory in Cambridge, UK. It included the creation of the Trillium user-interface design environment that enables designers to add design abstractions, Rooms (with Stuart Card) that moves beyond the desktop metaphor for managing window-based workstation tools, and Buttons that encapsulate actions for use, modification and sharing. His contribution to the book was written while Austin was an interface designer for the industrial design form of FitchRichardsonSmith of Columbus, Ohio, USA. Austin is co-chair of the ACM Special Interest Group on Human-Computer Interaction (SIGCHI). He has three children and lives with his wife in Palo Alto.

Berit Holmqvist is associate professor at the Institute of Information and Media Science, Aarhus University, Denmark, where she teaches Linguistics and esthetics, Design of user interfaces and Communication in organizations. Her Ph.D. thesis was on language use in a workplace, using empirical material from her studies at the Postal Giro in Stockholm. Her background is purely linguistic. Until four years ago she worked at the Department of Nordic Languages, University of Stockholm, teaching grammar and sociolinguistics. By chance she got in touch with some of the other authors in this book and her interest for the interplay between computers and communication in workplaces was born. She packed bags for herself and two daughters and moved to Aarhus where she still lives.

Morten Kyng is associate professor at the Computer Science Department, Aarhus University, Denmark. He teaches system development and does research on tools and techniques for the cooperative design of computer systems. His work on how to increase end-user influence on the design and use of computer systems has included the development of several courses for shop stewards. His first book in English, *Computers and Democracy* (Avebury, 1987), edited together with Gro Bjerknes and Pelle Ehn, introduced the idea of designing for democracy and skill at the workplace. He lives with his wife and two sons in a collective community of 30 families just outside Aarhus.

Finn Kensing is an associate professor at the Computer Science Department, Roskilde University Centre, Denmark. He does research on theories of, and methods for, system development, the area in which he also teaches. His concern for socially useful systems has led to teaching shop stewards how to work for their interests regarding the development and use of computers at their work. He is an author of *Professional Systems Development— Experience, Ideas and Action* (Prentice-Hall, 1990). The book presents a framework for planning, carrying out and evaluating system development projects. It is a result of research involving practitioners from software houses and internal computer departments.

Kim Halskov Madsen is associate professor at the Institute of Information and Media Science, Aarhus University, Denmark. He teaches program development and does research on tools and techniques for cooperative design. In addition, he teaches courses for shop stewards. His doctoral thesis is entitled *Language Use and Design* (Aarhus University, 1988). Currently he is on leave to research at George Mason University in Washington, DC.

Jesper Strandgaard Pedersen is an assistant professor at the Institute of Organization and Industrial Sociology, Copenhagen Business School, Denmark. His research interests focus on organizational symbolism, organizational change and information technology. His specific research interests have concerned cultural transformation processes in information technology firms, occupational subcultures, and mergers and acquisitions. He has co-authored a book on organizational culture, *Organizational Culture in Theory and Practice* (Avebury, 1990).

Dan Sjögren is a consultant at the Swedish Agency for Administrative Development in Sweden. He has a background as a researcher in organizational development, with a focus on design. His work is now oriented toward design methodology for the development of professional roles and work organizations in the context of information technology. The methodology is a combination of development and education and is based on the principle of end user involvement in the design processes. The development of the design methodology involves collaboration with the Institute of Information and Media Science, Aarhus University. He lives with his wife and three sons in Gröndal, Stockholm.

Lucy Suchman, a social/cultural anthropologist trained at the University of California at Berkeley, heads the Work Practice and Technology Area at Xerox Palo Alto Research Center. Her Ph.D. dissertation, a critique of research in artificial intelligence and human-computer interaction, was published in 1987 by Cambridge University Press under the title *Plans and Situated Actions: the problems of human-machine communication*. For the last ten years her research interests have focussed on the relation between studies of work practice and the design of appropriate computer systems. She shares a house and garden with her colleague Brigitte Jordan in La Honda, California, and draws inspiration from regular visits to Aarhus University.

Randall Trigg, associate professor at the Institute of Information and Media Science, Aarhus University, Denmark, received his Ph.D. in computer science in 1983 from the University of Maryland. His subsequent work at Xerox Palo Alto Research Center focussed on the areas of hypertext/hypermedia, computer supported cooperative work, and the design of tailorable systems. Close collaborations with social scientists at PARC led to a hypermedia-based system for supporting the work of interaction analysts, and more generally, to an interest in using methods and perspectives from the social sciences to better understand and inform the system design process. Recent work with Susanne Bødker and Kaj Grønbæk at Aarhus University employs techniques from interaction analysis in an analysis of the cooperative prototyping process. Randy's interests include jazz piano and accordion music from the Balkans.

Eleanor Wynn is publisher of the journal *Information Technology & People* (previously *OFFICE: Technology & People*). She has been involved in the study of social issues in computing since 1976, when she was a graduate research intern at Xerox Palo Alto Research Center for the summer. Her dissertation, titled *Office Conversation as an Information Medium*, represented one of the first academic attempts to document the background processes that people perform—as a function of their basic social competence—on seemingly routine, but actually intensively negotiated and contextualized procedures in a clerical administrative setting. Her interests span a range of topics that involve how people are perceived and incorporated into a system design. As a student of social interaction, she is presently concerned with the social dynamics of the system design interview as they affect the information generated therein. She lives near Portland, Oregon (with a view of Mt. Hood), with her son and spousal person.

Editors Joan Greenbaum and Morten Kyng getting "feet-on experience" with a marriage of art and technology—metaphorically represented by the merging seas of Kattegat and Skagerak.

Index

Printed and bound by CPI Group (UK) Ltd, Croydon, CR0 4YY

17/10/2024

01775687-0019